Language and Scientific Research

Wenceslao J. Gonzalez
Editor

Language and Scientific Research

palgrave
macmillan

Editor
Wenceslao J. Gonzalez
Center for Research in Philosophy of Science and Technology
University of A Coruña
Ferrol, Spain

ISBN 978-3-030-60536-0   ISBN 978-3-030-60537-7 (eBook)
https://doi.org/10.1007/978-3-030-60537-7

This Palgrave Macmillan imprint is published by the registered company Springer Nature Switzerland AG.
The registered company address is: Gewerbestrasse 11, 6330 Cham, Switzerland

Contents

Notes on Contributors

Jesus Alcolea is Emeritus Professor of Logic and Philosophy of Science, Department of Philosophy, at the University of Valencia, where he has been Dean of the Faculty of Philosophy and Educational Sciences. He researched at Smith College (Northampton, MA) with Thomas Tymoczko. His publications include "Proof as a Way to Increase Understanding" (1996); "Demostración como comunicación" (1997); "Fallibilism and Realism in Lakatos" (1999); "Vigencia del Pensamiento filosófico-matemático de Imre Lakatos" (2001); "Los conceptos matemáticos en el Mundo 3 de Popper" (2004), "Ontological and Epistemological Problems of Mathematics" (2006); "Kitcher's Naturalistic Epistemology and Methodology of Mathematics" (2011); "Argumentation in Mathematics" (2013): "A favor de la distinción entre 'argumento' y 'argumentación'" (2015); and "How to be a Critical but Reasonable Debater" (2018).

Jeffrey Barrett is Chancellor's Professor at the University of California, Irvine. From 2009 to 2017 he was Editor-in-Chief of Philosophy of Science, the official journal of the American Society for the Philosophy of Science. He is an advisor to important scientific committees. He has a degree in Physics and has completed his Doctoral Thesis in Philosophy (Columbia University). He specialized in Philosophy of Physics. He has also published on general Philosophy of Science, Decision Theory, Game

Theory and Logic. His publications include *Quantum Mechanics of Minds and Worlds* (1999), "Are Our Best Physical Theories Probably and/or Approximately True?" (2003), "Approximate Truth and Descriptive Nestling" (2008), "Faithful Description and the Incommensurability of Evolved Languages" (2010), "On the Coevolution of Basic Arithmetic Language and Knowledge" (2013), "Rule-Following and the Evolution of Basic Concepts" (2014), "Truth and Probability in Evolutionary Games" (2016) and *The Conceptual Foundations of Quantum Mechanics* (2020).

Anastasios Brenner is Professor of Philosophy at the Université Paul Valéry in Montpellier and member of the French National Committee for History and Philosophy of Sciences (Academy of Sciences). He is specialized in Historical Epistemology and the History of Philosophy of Sciences. In addition to publications on Pierre Duhem (1990, 1992, 1997), he is co-editor of the book *French Studies in the Philosophy of Science* (2009). Dr. Brenner is author of multiple works on the French contribution to this field, such as *Les origines françaises de la philosophie des sciences* (2003) and "Epistemology Historicized: The French Tradition" (2014). His recent publications include "Evolving Realities: Scientific Prediction and Objectivity from the Perspective of Historical Epistemology" (2020).

Wenceslao J. Gonzalez is Professor of Logic and Philosophy of Science (University of A Coruña). He is a Full Member of the *Académie International de Philosophie des Sciences/International Academy for Philosophy of Sciences*. He has been a Team Leader of the European Science Foundation program entitled "The Philosophy of Science in a European Perspective" (2008–2013). He has been *Visiting Fellow* at the Center for Philosophy of Science (U. Pittsburgh) and a visiting researcher at the London School of Economics. He has been a member of the National Committee for Evaluation of the Scientific Activity (CNEAI) of Spain. He is the director of the Center for Research in Philosophy of Science and Technology at the University of A Coruña, where he organizes every year a conference on Contemporary Philosophy and Methodology of Science. His publications include monographs such as *Philosophico-*

Methodological Analysis of Prediction and its Role in Economics (2015) and the edition of 42 volumes on philosophy of science and technology, such as *New Approaches to Scientific Realism* (2020).

Amanda Guillan is a Member of the Center for Research in Philosophy of Science and Technology at the University of A Coruña and a Vocal of the Foundation Philosophy of Science and Technology. Dr. Guillan is the author of the book *Pragmatic Idealism and Scientific Prediction: A Philosophical System and its Approach to Prediction in Science* (Springer, 2017). In this regard, she has done several research stays at the University of Pittsburgh. She actively collaborates with the Research Group of Philosophy and Methodology of the Sciences of the Artificial at the University of A Coruña. where she is a member of the organizing committee the annual Conferences on Contemporary Philosophy and Methodology of Science. Dr. Guillan has published on "The Limits of Future Knowledge: An Analysis of Nicholas Rescher's Epistemological Approach" (2016) and "The Obstacles to Scientific Prediction: An Analysis of the Limits of Predictability from the Ontology of Science" (2016). Among her most recent publications are "Realistic Components in the Conception of Pragmatic Idealism: The Role of Objectivity and the Notion of 'Fact'" (2020) and "Methodological Incidence of the Realms of Reality: Prediction and Complexity" (2020).

Tarja Knuuttila, is Professor of Philosophy of Science at the University of Vienna. Previously, she was Associate Professor at the University of South Carolina, Department of Philosophy, and Academic Research Fellow at the Academy of Finland, University of Helsinki. She has been a visiting professor at the California Institute of Technology. Knuuttila served as an Editor-in-Chief of *Science and Technology Studies*. She has studied scientific representation, modeling and scientific inference, with a special emphasis on synthetic biology and interdisciplinary relations between different scientific disciplines. Her publications include "Scientific Representation, Reflexivity, and the Possibility of Constructive Realism" (2014) and "Models Templates within and Between Disciplines: From Magnets to Gases — and Socio-economic Systems" (2016). She is the co-author of "De-idealization — No Easy Reversals" (2019).

Ladislav Kvasz is Professor at the Charles University of Prague since 2010, in the Department of Mathematics and Mathematics Education. He is a member of the Institute of Philosophy at the Czech Academy of Sciences. He has carried out research stays at the Universities of Vienna, King's College London, Berkeley and Technical University of Berlin. Among his awards is the Fernando Gil International Prize for the Philosophy of Science (2011). He has published on "On Classification of Scientific Revolutions" (1999), "Galilean Physics in Light of Husserlian Phenomenology" (2002), "The Mathemasitisation of Nature and Newtonian Physics" (2005), *Patterns of Change: Linguistic Innovations in the Development of Classical Mathematics* (2008), "Kant's Philosophy of Geometry-On the Road to a Final Assessment" (2011), "Language and the Limits of Science" (2016), and "Mathematical Language and the Changing Concept of Physical Reality" (2020).

List of Figures

1

The Relevance of Language for Scientific Research

Wenceslao J. Gonzalez

1 Origin of the Philosophical Relevance of Language for Scientific Research

Both in philosophy, in general, and in philosophy of science, in particular, language is recognized as having a prominent role.[1] For philosophical thought, language holds a key position in contemporary philosophy, in a historical shift from the classical approach and modern thinking.[2] Thus,

[1] This preeminent role is now best appreciated in the sciences of the Internet, within the framework of the sciences of the artificial, where the role of language in Web science is central, as semantic web research has highlighted. Cf. Tiropanis et al. (2015); and Hendler and Hall (2016).

According to James Hendler, "the social nature of the Web 2.0 sites primarily allows linking between people, not content, thus creating large, and valuable, social networks, but with impoverished semantic value among the tagget content," Hendler and Golbeck (2008, p. 15).

[2] This relevance accentuates the semantic or the pragmatic trait according to the approach on the theory of meaning, which is the initial focus for characterizing scientific language. This leads to choices that give priority to the sense and reference of the words or to the meaning conceived as

W. J. Gonzalez (✉)
Center for Research in Philosophy of Science and Technology,
University of A Coruña, Ferrol, Spain
e-mail: wenceslao.gonzalez@udc.es

W. J. Gonzalez (ed.), *Language and Scientific Research*,
https://doi.org/10.1007/978-3-030-60537-7_1

in classical thought the preeminence was in metaphysics as *philosophia prima*, insofar as the ontological vision was key to philosophizing. Thereafter, in the modern conception the theory of knowledge predominated, either from the perspective of the primacy of my own case (in rationalism or empiricism) or from the approach of a transcendental subject (in Kantism). In the contemporary turn appears the emphasis on language, where the attention to the problems of meaning (sense, reference, etc.) are relevant at least since the works of Gottlob Frege,[3] who developed a philosophy of language (cf. Dummett ([1973] 1981), a conception of logic and a vision of mathematics (cf. Dummett 1991).

Insofar as Frege puts the philosophical emphasis on the problems of language through the theory of meaning, he introduces a shift in contemporary thinking that Michael Dummett considers equivalent to that carried out by René Descartes for modern philosophy.[4] This implies that Frege is especially important for the emergence of contemporary thought. The issues regarding language are certainly relevant to formal sciences (such as logic or mathematics, which he cultivated), but also are central for the empirical sciences (natural, social or artificial).

Historically, Frege contributes to the philosophical shift because he influences at least three leading trends of contemporary philosophy connected to the relevance of language, each of which has had its impact in rethinking science from new perspectives: (a) the phenomenological line, which influenced a version of existentialism and later on opened the way to hermeneutics; (b) the analytical way, which branched out into diverse routes over the years; and (c) the neopositivist approach, which gave rise to a logical-methodological trajectory for the study of science, which

---

use. In addition, this in turn can lead to giving primacy to the significance of the words rather than to the context or to holistic options, when the meaning is seen to depend on a certain set or whole.

A philosophical characterization of the theory of meaning that has been very influential is found in Dummett (1975) and Dummett (1976).

[3] Cf. Frege (1892). See also Frege (1918).

[4] Frege "starts from meaning by taking the theory of meaning as the only part of philosophy whose results do not depend upon those of any other part, but which underlies all the rest. By doing this, he effected a revolution in philosophy as great as the similar revolution previously effected by Descartes; and he was able to this even though there was only one other part of philosophy to which Frege applied the results he obtained in the theory of meaning. We can, therefore, date a whole epoch in philosophy as beginning with the work of Frege, just as we can do with Descartes." Dummett ([1973] 1981), p. 669.

through reformulations has had a wide repercussion over four decades (cf. Suppe 1974).

Frege is at the origin of a series of ideas that, centered on the role of language, influence prevailing currents of thought in the twentieth century, which started in Europe and have had repercussions worldwide: (i) Edmund Husserl's phenomenology (cf. Husserl 1901, 1902), which also influenced later hermeneutics, where attention was paid to the social sciences from the ideas of understanding, interpretation and application characterized by Hans Georg Gadamer;[5] (ii) the analytical philosophy of Bertrand Russell and Ludwig Wittgenstein, which contributed to rethinking logic and mathematics;[6] and (iii) the conception of logical neopositivism of authors of the Vienna Circle, such as Rudolf Carnap, who thought of formal logic as a key element for mathematics (cf. Carnap 1931) and for the ideal of a unified science and led a trend of great projection in philosophy and methodology of science.

For many years — in the trajectory followed by logical neopositivism, logical empiricism and the "received view" — [7] the analysis of the language of science from a logical key prevailed in the mainstream philosophy of science. However, since the publications of the second Wittgenstein, the analytical conception brought language to the forefront from the perspective of usage, focused on meaning as use (cf. Toulmin 1953, 1971). This influenced the vision of science as a human activity during the "historical turn," when Thomas Kuhn's develops — in his first approach — the historiographic interpretation of scientific change (cf. Kuhn [1962] 1970). But it also left its mark on the third stage of his philosophical-methodological conception, which highlights language precisely (cf. Kuhn [1983] 2000).[8]

---

[5] Cf. Gadamer (1960). See also Gadamer (1975).

[6] Besides the relationship between Frege and Wittgenstein in the initial phase of the *Tractatus Logico-Philosophicus*, there are also common points with the later period of *Philosophiche Untersuchungen*, cf. Dummett (1981b).

[7] The "received view" is an expression used by H. Putnam the same year as the publication of Kuhn's main book. Cf. Putnam (1962). With it, the historical conjuncture lived at that time by the methodology of science of verificationist inspiration is adequately reflected.

[8] On Kuhn's stages of philosophical-methodological evolution and the turns in the role of language, see Gonzalez (2004b).

Until the 1960s, including the main writings of Karl Popper published by then,[9] science had prevailed as "content" and logic was a key instrument for its analysis. Once that primacy was abandoned in favor of science as an "activity" during the historical turn, then the diversity of nuances — intellectual and social — that language can provide began to be highlighted. Thus, language is no longer conceived as a mere expressive instrumental sign, where what matters is clarity in the statements made, but as part of the various aspects, internal and external, of the activity of research. This dual feature based on activity was reinforced in the following decades.

A new panorama was opened up from two completely different routes. On the one hand, the philosophical research on language, mainly Dummett's investigations regarding Frege's logic and mathematics,[10] but also in Brouwer's intuitionist mathematics (cf. Dummett 1977); and, on the other hand, the new trends in philosophy of science, which since the 1980s have been articulated in four main lines of thought. [11] Each of these conceptions assumes, implicitly or explicitly, a characterization of scientific language (usually for basic science and occasionally for applied science and the application of science).

These four philosophico-methodological lines are: (i) naturalism, which is diversified in various orientations (cognitive, normative, etc.) and usually competes in the social sciences with interpretivism;[12] (ii) the social concern on science (the strong program of Edinburgh, the empirical program of relativism, the ethnomethodology, etc.); (iii) scientific realism with its various options (structural realism, critical realism, etc.);[13]

---

[9] Cf. Popper (1935, 1945a, 1945b, 1957). See Gonzalez (2004a).

[10] Other authors paid attention to Frege over the years, which led to the publication of Dummett (1981a). In turn, within the philosophy of language, in general, and in the sphere of reference theory, in particular, there was an abundant number of publications and in various philosophical directions. See, in this regard, Gonzalez (1986b).

[11] Cf. Gonzalez (2006a), especially, pp. 1–19.

[12] Cf. Salmon (1992), especially, pp. 408–410.

[13] "These contemporary versions of scientific realisms include the following philosophical versions, among others: structural realism, critical realism, referential realism, entity realism, instrumental realism, socially embedded realism, constructive realism, some versions of scientific perspectivism (or perspectivalism), dispositional realism, convergent realism, pragmatic realism, selective realism, minimal realism, and the so-called 'preservative realism'." Gonzalez (2020a), p. 4. The main features of these various orientations can be found in Gonzalez (2020a), pp. 6–16.

and (iv) approaches based on probability theories, mainly the Bayesian interpretation of probability. This meant that there were internal and external aspects of scientific activity which, until then, had not been explicitly addressed.

Also in these last decades of the twentieth century, with pioneers at the end of the 1970s (like Patrick Suppes [1981, PSA 1978] and Nicholas Rescher [1977]), there are two approaches that are nowadays particularly influential regarding scientific processes: methodological pluralism and methodological pragmatism (cf. Gonzalez 2020c). (I) Methodological pluralism brings about variation of research according to the levels of reality (micro, meso, and macro) (cf. Gonzalez 2020d). In principle, this may lead to variations of scientific language — including sense and reference — following the scale of research, which are commonly more in tune with a pragmatic approach to meaning. (II) Methodological pragmatism depends on the specific context of the goals sought in the research made and the effectiveness in achieving them. Thus, scientific language takes on a pragmatic character in its meaning.

2 Language is One of the Constitutive Elements of Science

From a thematic viewpoint, language is one of the constitutive elements of science.[14] Both the sense and reference of its words and the semantic content expressed through its statements are a fundamental part of science and a key way to access its contributions. The rigor that accompanies scientific language is greater than that required of ordinary language and is perfected with new terms (boson, supernova, Internet, etc.), terminological extensions (digital economics, education online, …), variations in meaning over time (atom, evolution, web, …), etc., since science is also characterized by its historicity. Thus, in order to grasp the scientific contributions of the past, it is necessary to master the terminology of the

[14] The role of language, structure, knowledge, methods, activity, ends and values as constitutive elements of science is set out in Gonzalez (2013b), especially, pp. 15–17.

period under study and be able to evaluate its validity by mastering the contributions of our time.

As a constituent element of science, language accompanies the other elements that configure science: the structure in which scientific theories are articulated, scientific knowledge, research methods, scientific activity, scientific aims and the values of science (among them, ethical values).[15] But we come to know and master these constitutive elements of science by possessing the appropriate language or creating a new terms and statements to access the new realities, as is usually the case in the sciences of the artificial and as is seen in the sciences of the Internet (cf. Gonzalez and Arrojo 2019).

Through language we can access the internal structure of science. We can articulate it in macro-theoretical frameworks, theories, models, hypotheses and contrasting processes (via observation or experimentation in empirical sciences based on the available evidence). Also through language we can know a remote past scientifically, be able to reflect a lived present and be capable to anticipate a possible future. Through language we can know what we know, know what we do not know and be aware of the unknown to be known (i.e., some possible things that, to date, we have not known).

Furthermore, the methodological processes to advance scientific knowledge, either with initial procedures or with well-established scientific methods, require specific mastery of the appropriate terminology. It is through language that we can distinguish between data, information and knowledge, because we can differentiate between knowledge that organizes information (categorizes it according to criteria) and mere available information (a content accessible on a computer) if the researcher or the center of research has the necessary language skills for this purpose. In this respect, the data as an isolated piece of information is not intelligible for scientific purposes until it is arranged into information and categorized to be considered scientific knowledge.

Moreover, the scientific activity carried out by agents (individuals, groups, etc.) and organizations (public or private, national or

[15] All of them are particularly important for having a proper analysis of central scientific issues, such as scientific prediction. See Gonzalez (2015a), pp. 11–13

international, etc.) requires mastery of language. This language might be of different kinds. It can be specific to a given discipline (physics, chemistry, economics, pharmacology, etc.) or it may be an interdisciplinary lexicon used in some cases (like biochemistry, psychopedagogy, etc.).[16] In addition, the language can be a terminology accessible in other methodological options, such as a multidisciplinary, a transdisciplinary or a crossdisciplinary manner. But the requirement for scientific undertaking is the mastery of the sense and reference of the terms and grasping the statements made with them.

However, within the ontology of science, language is not an end in itself, because language of scientific activity has to deal with what *is or might be* real (natural, social or artificial) and has to give an *image* of that extramental reality (besides helping us to know the mental processes). In this respect, the first thing is to be able to *identify* the real, something that already exists or may come to exist, for which the language has to select an object or a process among a series of them (like Covid-19 when there are other coronaviruses). Then, after a change of space or time, it has to be possible to *reidentify* the previously identified object or process.

Being able to identify and re-identify something real is central to the theory of reference. This aspect is a very important part of the philosophy of language (cf. Gonzalez 1986a), before being a key ingredient of the semantics of science.[17] Through reference, a relationship is established between a word and an object or process, through the mediation of sense. This relationship is the semantic value associated with the term. At the same time, there is a real reference (in the past or at present) or a possible reference (in the future, be it ontological, epistemological or heuristic). It is the real referent what allows to associate the truth value to the statement that is related to that object or process. But it is also possible that, if there is not a referent designated by the subject of the statement, it can be dictated that the statement is not true or false (like some cosmological statements, on the basis of current knowledge).

It can then be considered that scientific language, insofar as it is endowed with a well-defined sense and reference, entails an ontological

[16] On interdisciplinarity, see Niiniluoto (2020).

[17] This issue is discussed in detail in Gonzalez (2021).

commitment. Thus, as soon as we say 'the electron exists' or 'there are gravitational waves,' we are admitting that there is something extramental that is real,[18] which is the electron or the waves predicted by Albert Einstein, whose existence as a referent — as an object or as a process — allows the statement to be true. When it comes to empirical sciences, the natural, social or artificial reference is assumed to be the basis for being able to say something true about the world.[19]

Concerning scientific prediction, which anticipates the possible future, the reference is needed as a relationship to something that is not yet or is not yet known. In this case, the reference works as a semantic value that allows the predictive statement to be correct. It is then about a reference that is possible (ontological prediction), which already exists but has not come to be known (epistemological prediction) or that enters within the range of the possible that can come to exist (heuristic prediction). This heuristic option is important in cases such as a pandemic, when we analyze a series of possible routes and we eventually state that 'the Covid-19 virus will not infect anyone in month x' or when we predict 'a new coronavirus may emerge in x years.'

We access the values of science through language. What is worthy of merit or what is considered demerit requires the use of terms to recognize them. We can use language to value means and ends of scientific activity (cf. Gonzalez 2013a). This concerns both the internal values (cognitive, methodological, etc.) and the external values (social, economic, cultural, ecological, etc.) of scientific activity (cf. Gonzalez 2008a). These values can be objective, if they respond to human needs, or they can be optional (cf. Rescher 1999). But, in order to function as actual values, they must be accepted by the agents and organizations that do science. This implies a component of historicity in the recognition of values, which is common regarding optional values, but also because some values that were initially external (such as sustainability) can become internal (for selection of ends and means).

---

[18] Ian Hacking has insisted that not everything is constructed. See Hacking (1999).

[19] The referent is then part of what we consider to be a "fact," so without a referent we can hardly have anything that we can call "fact." Regarding what is a *fact*, Peter F. Strawson wrote: "facts are what statements (when true) state." Strawson (1950), p. 136.

Within the values that go with scientific research are the ethical values (cf. Gonzalez 1999). These values can be endogenous, if they are part of the scientific activity itself, insofar as it is a free human activity with ends and means chosen by agents (individuals, groups, etc.).[20] Meanwhile, ethical values can be exogenous, if they depend on the environment (social, economic, institutional, etc.) where the scientific activity takes place. The language allows us to distinguish the types of ethical values and their relation to scientific research as human activity. From an endogenous viewpoint, ethical values can have a role about the diverse steps of scientific research: 1) the aims of research, 2) the processes for carrying it out, 3) the results expected, and 4) the consequences arising from those results. From an exogenous point of view, ethical values can have a role to play regarding scientific activity in its relation to other human activities.[21]

3 Main Roles of Language in the Configuration of Science

Among the tasks of language in science there are two that are especially important for scientific activity. First, it contributes to *establishing scientific thinking*. This is the case in at least four ways: (a) with adapted terms that are initially based on natural language (force, mass, etc.); (b) terms that were available in philosophical approaches (atom, mind, etc.); (c) terms introduced because of scientific discoveries (fermion, boson, etc.), which can lead to the replacement of an old term that is no longer acceptable (phlogiston) in favor of a new term (oxygen); and (d) terms that come through the use of a formal language, such as mathematics (topology, fractal, etc.). Thus, language shapes how scientists conceive scientific activity (problems, models and the ways to contrast scientific knowledge) and how society accesses scientific knowledge. Second, language has a *heuristic function* for scientific activity, insofar as it allows us to explore new possibilities, create new forms of expression for possible phenomena

[20] On collective morality, see Rescher (2003).

[21] As science has a role in shaping technology, its values also have a role for values in technology, cf. Gonzalez (2015b).

(in the short, middle and long run) as well as at different levels of reality (micro, meso and macro).

Regarding the first task, it happens that the *mode of expression* of scientific thought is an essential part of that same thought. But language is not a mere instrumental sign that "covers," "vehicles" or "transmits" a content: language shapes the scientific thought since it is an active mediator, gives it a specific form. In this respect, language commonly has a contextual component and a pragmatic factor, since that configuration looks towards some sought goals and performs this role within a specific context. De facto, language does so in various ways, including the use of specialized language based on natural language, which gives rise to new terms, a change in the meaning of existing ones, or the discarding of words that are now considered to have no real reference. Furthermore, it can do so by means of mathematical language, which might be an indispensable part of scientific thought (see Hendry 2012; and Gillies 2012).

Overall, language contributes to configurate the structure and dynamics of scientific research, insofar as it affects internal and external aspects of scientific activity: (i) the work of the agents who carry out research in a center (public or private, national or international, etc.); (ii) their relations with other colleagues, which require adequate communication, which is essential for having a scientific community; (iii) their connections with the institutions, which should understand their task (the assumptions, contents and limits of their research); and (iv) their links with society, which in moral and economic ways should support their efforts to achieve new discoveries, to solve practical problems and to apply the solutions in new contexts.

Certainly, language opens new possibilities insofar as it also has a *heuristic function* for scientific activity. This can be done through changes in the natural-based language, as Charles Darwin did with respect to the use of "evolution," which opened up a whole panorama of discoveries, or through mathematical language, so that it generates new models for facing various problems, such as climate change, the spread of pandemics provoked by Covid-19 virus or the migration of inhabitants in countries in conflict.

This change of sense and reference of "evolution" introduced by Darwin, in addition to influencing biology in its various scientific branches (like botany, zoology, ecology, etc.) and social sciences (like psychology, sociology, anthropology, etc.), has affected philosophy from several angles: (a) in philosophical questions of the systematic branch, such as the theory of knowledge or metaphysics; and (b) in philosophy of science, both in the general orientation (epistemological, methodological, ontological, etc.) and in the special one (philosophy of biology, philosophy of economics, etc.) (cf. Gonzalez 2008b). It has also had a cultural impact, one of the facets of which is precisely the public's perception of science.

Many researchers, especially in some disciplines (physics, economics, epidemiology, etc.), highlight the heuristic use of mathematical language. This is the case of Herbert Simon:[22] "For me, mathematics has always been a language of thought. (…). When I am working on a problem, I am sure that I do not usually think in words, but in terms of a more abstract representation that is perhaps partially pictorial or diagrammatic and partially symbolic. Mathematics — this sort of nonverbal thinking — is my language of discovery. It is the tool I use to arrive at new ideas. This kind of mathematics is relatively unrigorous, loose, heuristic. Solutions reached with its help have be checked for correctness. It is physicists' mathematics or engineers' mathematics rather than mathematicians' mathematics." (Simon 1991, 106–107).

Simon compares his view of mathematics as a *language of discovery* with another Nobel Prize winner, who puts the emphasis on the other pole, which is mathematics as a *language of proof*: "For Tjalling Koopmans, it appeared, mathematics was a language of proof. It was a safeguard to guarantee that conclusions were correct, that they could be derived rigously. Rigor was essential. (I have heard the same views, in even more extreme form, expressed by Gerard Debreu; and Kenneth Arrow seems mainly to share them.) I could never persuade Tjalling that ideas have to be arrived at before their correctness can be guaranteed, and that the logic of discovery is quite different from the logic of verification. I am sorry

[22] The use of mathematics for the heuristic function through predictions is what focused attention during the pandemic generated by the Covid-19 virus. Various mathematical models have been used to anticipate the possible future course of the disease at a general level and in each country. They have been used by the World Health Organization to make recommendations and by health authorities in each nation to make decisions.

that he did not live to read and comment upon my recent work on the logic of scientific discovery. Perhaps we could have built a bridge across what seemed a great gulf that separated out attitudes toward mathematics. It is his view, of course, that prevails in economics today, and to my mind it is a great pity for economics and the world that it does." (Simon 1991, 107).

Obviously, the relationship between mathematical language and empirical reality (whether natural, social or artificial) is a matter of constant reflection. [23] For Margaret Morrison, "abstract mathematical models can often furnish the kind of physical information that is not available using methods of empirical investigation" (Morrison 2015, 50). She wants to "address the question of what actually constitutes a mathematical explanation of a physical fact" (Morrison 2015, 50). This question is "important not only because much of the contemporary science, especially physics, is expressed in mathematical language but also because new phenomena are sometimes predicted on the basis of mathematics alone (e.g. the π meson)" (Morrison 2015, 51).

4 Language in Basic Science, Applied Science and Application of Science

If we accept the existence of a scientific diversity, which is expressed through relevant philosophical-methodological variations in scientific research that prevent methodological universalism and question methodological imperialism (cf. Gonzalez 2012), then it is easier to recognize the existence of differences between basic science, applied science and application of science. In this ambit, it seems clear that language allows differences between basic science, applied science and application of science to be shaped. De facto, they use a variety of scientific statements, which can

[23] The philosophical status of mathematical language has also been discussed. In this regard, there are also various philosophical orientations. See, in this regard, the text presented in the 2008 Biennial meeting of the Philosophy of Science Association: Psillos (2010). A different approach can be found in Azzouni and Bueno (2016).

It seems to me important to consider that mathematics is also a human activity, which has philosophical consequences. See Gonzalez (1991).

be diversified respectively into three main groups: (i) explanatory and predictive statements; (ii) predictive and prescriptive statements; and (iii) statements oriented towards differentiated contexts of use.

(I) *Basic science*, which has as its primary task the rigorous expansion of available knowledge, is oriented towards explanation and prediction. The language of explanation, as Wesley Salmon has put it, has to do primarily with the answer to the question of *why* (cf. Salmon 1990). This inquiry was usually aimed at finding out why something occurs (it was usually looking for a cause in his approach) (cf. Salmon 1998), rather than merely knowing *that* something is happening. The statements that answer the question of *why* can lead us to explanatory knowledge (cf. Salmon 1990, 3), so they allow us to do science, while the statements of knowledge *that*, if they are merely descriptive, do not guarantee scientific knowledge.

But the reality is often multifaceted (especially in complex systems) and the statements that answer the question of why may follow very different paths. Thus, that answer can be diversified into different types of scientific explanations. Among them are nomological-deductive, probabilistic-inductive, functional or teleological, and genetic or historical explanations (cf. Gonzalez 2002). They lead to various methodological directions, which rules out a single or uniform way of explaining scientifically (even within the natural sciences). Moreover, each of these lines of research has its own style of statement, which is related to how the object is studied and how the problem is posed.[24]

Explicative language is usually about what has been (near and distant past) or about something that now is (our present), while predictive language deals with that which does not yet exist (ontological prediction), that which is not yet known (epistemological prediction), or that which can take various forms according to the routes followed (heuristic prediction). This anticipation of the expected (reality, knowledge or range of

---

[24] The idea of diversity in scientific explanations is already present in the influential book of Nagel (1961). Although, over the years, Salmon made various proposals for scientific explanations, they usually revolved around three elements: preference for the causal explanation over other types of explanations, emphasis on the role of probability, and recognition of the presence of pragmatic elements that modulate explanations. His final proposals can be found in Salmon (2002a); and Salmon (2002b)

possibilities), based on known variables, made by scientific prediction can be in the immediate, short, medium, long and very long run.[25]

Besides the aspect of anticipation and the type of object studied, the degree of control of known variables matters greatly. Thus, prediction — understood in *lato sensu* — gives rise to statements of three different types. These are foresight, prediction (in *stricto sensu*) and forecasting, where *foresight* involves statements of a clear control of variables involved; *prediction* states the situation of a variable within a period of time, when it happens that the variable is not really under our control; and *forecast* leads to predictive statements with a margin of error associated. These three cases are different from planning, which is a concept of action and this teleological orientation assumes foresights, predictions and forecasts. Planning uses them to obtain certain ends in a defined spatial and temporal framework.[26]

Prediction as a *scientific test* is the methodological task it has in basic science, where it is a question of checking the scientific validity of statements that anticipate the future (ontological, epistemological or heuristic). This is done in order to conclude by dictating whether the theories, models or hypotheses in which the prediction is framed can be sustained or have to be discarded or be replaced by others. Thus, prediction is related to novel facts,[27] which are of different kinds. The researcher knows or should know that there are various types of novelty and that prediction can be taken as a criterion for appraisal according to these various types of novelty (theoretical, empirical or heuristic) (cf. Gonzalez 2015a, 110–119). In some disciplines, such as economics, the role of prediction as a scientific test has had an undoubted philosophical-methodological prominence (cf. Gonzalez 2006b).

(II) *Applied science* uses prediction in a different way, because the aim of this type of research is the resolution of concrete problems (cf. Niiniluoto 1993, 1995). In this case, predictive language is not directly oriented towards the validity of what has been stated. In applied science,

[25] Cf. Gonzalez (2015a), pp. 66, 192, 219, and 251.

[26] A detailed analysis of the distinction between foresight, prediction, forecast and planning is in Gonzalez (2015a), pp. 68–72.

[27] Imre Lakatos insisted on this point. Cf. Lakatos (1970). See, in this regard, Gonzalez (2001).

prediction has the role of *a guide to action* which, by anticipating the possible future, establishes the feasibility or otherwise of prescription in order to solve the specific problem posed (natural, social or artificial). Moreover, in applied research scientific models are not "descriptive" (strictly speaking, they are explanatory, predictive or both), because they are *prescriptive* (related to patterns of action) (cf. Simon [1990] 1997).

Prescription includes, implicitly or explicitly, the role of values in the practical models, to evaluate the positive and negative sides of each course of action. This should help to dictate which prescription among the possible ones is the most appropriate. This feature is particularly important in sciences such as economics, where prescriptions have clear social implications.[28] Thus the type of language used in applied science is different from basic science. This characteristic is better appreciated when it comes to design sciences (cf. Gonzalez 2008c), within the framework of sciences of the artificial. Because then *synthesis* predominates over analysis, insofar as the design seeks something artificial with some desired properties (cf. Simon 1996, 4). The designer, as Simon has pointed out, "is concerned with how things *ought* to be — how to be in order to *attain goals*, and to *function*" (Simon 1996, 4–5).

As applied science uses predictions and prescriptions, where there is a cadence of aims, processes and results (leading to consequences, which might be unexpected), the scientific language used in applied research is different from that of explaining and predicting of basic research. From this point of view, pharmacology does not investigate in the same way as genetics or molecular biology, although it needs the knowledge of genetics or molecular biology to make some its predictions about the ability to cure or mitigate a disease that can have some active ingredients of a medicine. Meanwhile, the prescription for using a medicine can consider a number of goals and a scale of priorities, based on the direct or indirect effects anticipated by the pharmacological predictions.

Conceptually, there is a distinction between two uses of language in science: (a) the scientific language about what we can observe or experiment,[29] but we cannot have an intervention to achieve concrete

[28] See Sen (1986); especially, p. 3; and Gonzalez (1998a).

[29] On the various options for observation and experimentation, see Gonzalez (2010).

aims of a practical character (as is in the case with cosmology or in the study of the earth's geological core), and (b) the language used when we may have to prescribe a course of action (as in climate change to lower the level of carbon dioxide of human origin, in ecology to avoid the disappearance of species, in the reduction of unemployment or in the development of new versions of the Web). At the same time, it can be clearly maintained that there may be a *bidirectional relationship* between basic research and applied research. This kind of operative intertwining can be found in cases such as the recent painstaking search for a vaccine for Covid-19[30] or the endeavor to achieve viable solutions to prevent a rise in sea temperature worldwide.

Following the current philosophical-methodological preference in favor of pluralism and the use of pragmatic criteria for research, what is discarded is the vision of scientific language as a kind of logical-methodological edifice based on "fundamental propositions" or "protocolary propositions."[31] Also, criticized is the epistemological-methodological idea of the basic science that provides the "fundamental" branch of the discipline (in physics, chemistry, biology, psychology, etc.) and which, on this basis, the rest of the building of this scientific knowledge is constructed.[32] Furthermore, highlighting applied science as having its own philosophical-methodological characteristics, different and complementary to basic science, contributes to seeing scientific activity as having different statements, depending on the purposes of the research.

(III) *Application of science* is a third area of use of scientific language. Because applying science is different from applied science, insofar as it seeks the practical use of science according to different contexts of use. Thus, if applied science offers a pattern to give solutions to the problems that arise (as happens in teaching in schools of medicine), the application

[30] "It is a testament to the machinery of science that so much has been learned about covid-19 so rapidly. Since January the number of publications has been doubling every 14 days, reaching 1363 in the past week alone. They have covered everything from the genetics of the virus that causes the disease to computer models of its spread and the scope for vaccines and treatments." The Economist (2020).

[31] Current positions on this issue can be found in Gonzalez (2020c).

[32] That science investigates according to a methodological diversity and according to scales of reality, with epistemological differences according to levels is increasingly assumed. Cf. Gonzalez (2020d).

of science is the translation of those solutions according to the corresponding contexts (as happens in hospitals, which apply the therapies to specific people with medical histories that may be completely different). This third type of language is the closest to the main problems of nature, society or the sphere of the artificial.

There is also a *bidirectional relationship* between the application of science and applied science. This involves (i) the use of scientific knowledge in well-defined contexts (in space and time), which provides guidance on possible uses of solutions considered to be viable, and (ii) the results actually obtained to solve the specific problems posed, which can contribute to shaping the appropriate solutions in applied science. Again, there are predictive and prescriptive statements, since in the application of science it is necessary to anticipate the possible future (like how the infection of a patient with coronavirus may evolve) in order to provide a solution in this defined field (like therapy for a patient with a known medical history).[33]

Nevertheless, the application of science is not the same as *public policy* or the *administrative management* of the experts' recommendations. This distinction should be highlighted due to what has happened in many countries, especially during the pandemic generated by Covid-19. The application of science is the responsibility of scientists, while public action or health policy is the task of each country's health managers. It must also be recognized that there may be a wide gap between the contributions of scientists and public perception (cf. Melo-Martin and Intemann 2018), both at the general level and in each country. Moreover, there may be a phenomenon of mistrust towards the experts, which may come from governments, the media or the citizens of each territory.

[33] To date, treatment of Covid-19 has often been completely individualized, if not purely ad hoc through trial and error, due to the absence of previous well-founded studies offering well-contrasted solutions to the disease.

5 Scientific Language Cannot be Reduced to Structural Components

Scientific language cannot be reduced to structural components (macro-theoretical frameworks, theories, models, hypotheses, etc.), because it encompasses dynamic components of science as well. The reference of the terms used in the structural components is undoubtedly important. In this respect, it is assumed that the reference itself (i.e., the referent pointed out by the term used) is, in principle, intelligible as are its properties and internal configuration. Consequently, it is therefore particularly strange to maintain that "the epithet 'intelligible' applies to theories, not to phenomena."[34]

According to this view, the reference in science seems to be opaque, since this gives the impression of implying that there is no genuine understanding of the phenomena (i.e., actual intelligibility), insofar as de Regt uses "the terms 'intelligibility' and 'intelligible' only in connection with the understanding of theories" (de Regt 2017, 12, note 8). In this regard, his definition of *intelligibility* does not look at the phenomena (i.e., the referent) but only towards the theory: intelligibility is "the value that scientists attribute to the cluster of qualities of a theory (in one or more of its representations) that facilitate the use of the theory" (de Regt 2017, 40).

Very often the references change over time and this aspect could be intelligible too for sciences. Their variability through time is a question that has to be reflected using the language without losing scientific objectivity. Also these dynamic components often cannot be condensed merely in terms of processes and evolution, due to the complexity of many phenomena considered and the kind of variations possible, such as "scientific revolutions."[35] Commonly, at least the social sciences and the science of the artificial do research on phenomena characterized by historicity:[36]

[34] de Regt (2017), p. 12; see also (2017), pp. 45 and 88.

[35] There has been a very intense debate on the existence and characteristics of the scientific revolutions. After the very influential book by Thomas Kuhn on *The Structure of Scientific Revolutions*, an important contribution was made in Thagard (1992). See, in this regard, Gonzalez (2011a).

[36] Cf. Gonzalez (2011b, 2013c).

(a) There is a variability of phenomena (especially social and artificial) that contributes to the historicity of language, which requires appropriate terminology for new realities available[37] or for reinterpreting those already known. (b) This linguistic historicity connects with the relationship between the researcher and his or her environment, since this link is mediated by the factor of change over time and by the progress of research itself.[38] (c) The researcher's own social and institutional context (his or her career, relations with colleagues and links with organizations) is modulated by the historicity expressed in new terms, nuances in existing terms and the abandonment of terms considered obsolete.

Historicity, for Thomas Kuhn in his famous first philosophical-methodological stage,[39] begins at the nexus between language and perception, which conceives in a relativistic way. This is because language is part of the paradigm (at least, it belongs to the theory) and the interpretation of perception depends on the conceptual schemes of the accepted paradigm, which are assumed when interpreting the experience of the world and vary radically by opting for an alternative paradigm in a scientific revolution. Thus, by focusing on cases from physics and chemistry, he maintains that in a scientific revolution "though the world does not change with a change of paradigm, the scientist afterward works in a different world" (Kuhn [1962] 1970, 121).

Kuhn's initial philosophico-methodological stage incorporates in fact a semantic holism regarding paradigms, so that by changing the semantic set for a new paradigm there is a radical change. In addition, there is a "Gestalt switch,"[40] i.e., a change in perception by looking at the same

[37] In the case of sciences of Internet the novelty is clear, cf. Hall et al. (2016). The design of the network itself is clearly new, cf. Clark (2018). Overall, it can be said that we are in a new historical stage, which Luciano Floridi calls "hyperhistory," cf. Floridi (2014).

[38] "As the deluge of work on covid-19 has shown, fast, free-flowing scientific information is vital for progress. The virus has changed the way scientists do their work and talk to each other, we hope for good." The Economist (2020).

[39] At least three philosophical-methodological stages can be distinguished in Kuhn's publications, cf. Gonzalez (2004b), especially, pp. 48–66.

[40] As a result of the criticism received in the first stage, Kuhn introduced a series of relevant philosophical-methodological changes in the second period. Among them was the prominent role of exemplars, as characteristic solutions to problems posed and accepted as such by the scientific community. In this regard, Kuhn's second stage — with the exemplars as a route to learn scientific theories — has been associated with concept characterizations within the framework of cognitive

phenomena: "rather than being an interpreter, the scientist who embraces a new paradigm is like the man wearing inverting lenses. Confronting the same constellation of objects as before and knowing that he [or she] does so, he [or she] nevertheless finds them transform through and through in many of their details" ([1962] 1970, 122). This position is clearly the opposite to the approach of reports on "the given," which dominated the scene for decades during the influence of the logical-methodological trajectory that began with the Vienna Circle. Thus, if the relevance of historicity is accepted and scientific revolutions are possible, then scientific language cannot be restricted to informing the merely "given." [41]

Before and after Kuhn's dynamic view of scientific perception as shape or figure shifting, where structurally observation and experimentation are ladened theoretically,[42] there have been other influential positions on how to *conceive perception* — structurally or dynamically — when observing or experimenting in empirical sciences (natural, social or artificial) and to what extent it is *mediated linguistically*. This issue has had very diverse responses: from philosophical positions with modern roots (rationalist, empiricist, Kantian, idealist, etc.) to approaches that are directly based on contemporary philosophical conceptions, such as analytical philosophy.[43]

De facto, the philosophy of science in recent decades has diversified in many directions in dealing with the phenomenon and the role of language. Thus, although the semantics of science is not usually the express starting point of the various recent philosophical trends, these often share the idea of the observation being theoretically laden. This happens, in principle, in the different varieties of scientific realism, in positions of naturalism, in conceptions that emphasize the social dimension of

---

psychology. It is clear that the first stage was under the influence of the Gestalt psychological school he knew. Later, that school moved towards a characterization of concepts more in tune with classical positions, where concepts represent features that are typical of a defined class of objects. Cf. Andersen et al. (2006). A critical analysis of the book can be found at Thagard (2009).

[41] Cf. Kuhn ([1962] 1970), p. 127.

[42] "Paradigm changes do cause scientists to see the world of their research-engagement differently," Kuhn ([1962] 1970), p. 111.

[43] Analytical philosophers who have dealt with perception include Gottlob Frege and Peter F. Strawson. For the former, see the chapter "Frege on Perception," in Dummett (1993 [reprinted 1998]), pp. 84–98. For the second, see Strawson (1961) and Strawson (1979).

science, in different lines of pragmatism, in pluralism that rules out a single epistemological reading of what is perceived, etc.

For basic science, the role of language is important for the structural factor of determining what it really is to "perceive" scientifically (and, therefore, what it is to "see" in current science or getting something relevant through the other senses, particularly hearing) when observations and experiments are modulated by the instruments available. Furthermore, the language used is key for the dynamic aspects in this territory: (i) when it comes to changing theory over time and (ii) regarding how scientific rationality is maintained with changes in language (variations in terms, new interpretations of existing ones, completely new terms, etc.). These issues are central when dealing with *conceptual revolutions* in science, which allow us to understand the intertwining of the history of science and the philosophy of science in the wake of historical cases, such as the abandonment of "phlogiston" in favor of "oxygen," which led to A. Lavoisier's chemical revolution.[44]

Unquestionably, the act of perceiving scientifically (while looking through the electron microscope, using the space telescope or doing a positron emission tomography) requires a conceptual component, which in semantic terms is a link to a sense — a content that aspires to be objective — and a reference, which is related to some object or process. To perceive scientifically is an act mediated conceptually and associated with a network of meaning. This is required so that the data appearing in the observation or experimentation can give rise to an informative scientific statement. Based on the informative content scientific, knowledge can be derived, insofar as the information received can be organized, categorized or classified in order to be genuine knowledge (and, therefore, to give rise to scientifically tested content).

Consequently, there is then a "theory-laden" in perception, which can change in terms of historicity, insofar as new conceptual content emerges to focus on what is perceived. This can be better understood if the polyhedral character of reality is taken into account, where objects or

[44] The chemical revolution receives special attention in Thagard's perspective on conceptual change. Cf. Thagard (1992), pp. 34–61; especially, pp. 39–47. An analysis of Thagard's conceptual revolutions and the need for new aspects can be found in Gonzalez (2011a), pp. 15–21. A review of Thagard's book is available in Gonzalez (1996).

processes present diverse aspects.[45] In this regard, to be theory-laden presupposes that perceiving (observing or experimenting) is not a passive act but an active one. Moreover, to perceive scientifically is understood, in principle, as an intentional act oriented towards an end, based on the search for a resolution of a problem posed, and all mediated linguistically.

In turn, the reality that is perceived, especially in the social area and in the sphere of the artificial, is subject to historicity. Thus, the change in reality favors a new conceptual content for the language used, [46] which captures new aspects of the object (properties, structural elements, etc.) or new processes. So there are two elements involved here: first, the conceptual content itself and, second, the field covered by the concept used. Both aspects — intensional and extensional — are relevant to comparing what two scientists actually perceive when they observe something or when they do an experiment.

By moving from the use of terms — with their sense — to identifying what is perceived to statements that seek to express a "fact" (i.e., a phenomenon or event with its defined characteristics), which can be identified and re-identified in the appropriate conditions, then the theory-laden component is expanded. In principle, underneath can be macro-theoretical frameworks, theories, models or hypotheses that address the observed or the experimented reality in relation to a scientific problem presented. Scientific activity has intentionality — it looks towards a goal — and, in this search, it wants efficiency in the resolution of the problem at hand. Therefore, not just any conceptual content is valid: we must aspire to *objectivity*,[47] which can lead to true knowledge. The first step in this scientific research is then semantic, whereas the second step is epistemological with methodological repercussion.

[45] For processes from an ontological viewpoint, see Rescher (1996).

[46] This is the origin of the friendly controversy with Peter Strawson on the characteristics of the concepts. It started with Gonzalez (1998b). The matter went on with his answer: Strawson (1998). It was then completed in another subsequent paper: Gonzalez (2003).

[47] The central role of objectivity in the search for truth in science is emphasized in Gonzalez (2020b).

6 Language of Scientific Activity and Language of Other Activities Connected to It

Commonly, the dimension of language that is highlighted is that it is present in *scientific activity*, with its ends, means and results (from which consequences are derived). In addition, we should consider the language of *this activity as connected with other human activities* in social life. These two aspects involve at least the following features: (i) The fact that science is a human activity mediated linguistically emphasizes at least two things: (a) that it is *our activity* — human agents do collaborative actions using language to do scientific research — and (b) that when we do science we also *do things with words*.[48] (ii) The linguistic dimension of science as a human activity is intertwined with other human activities but it is different from other linguistic codes. Thus, scientific language is different and complementary to the use of ordinary language and it is also diverse from the specific technological language, although science certainly contributes to technology in a direct way.

Subsequent to the first line, the human character of science is highlighted, even though more and more tools are used (which, in turn, are the fruit of human design). Furthermore, some aspects of the scientific activity are stressed insofar as we do things with scientific words and statements. Thus, in addition to using words for explanatory, predictive, prescriptive and scientific knowledge application tasks, we also perform acts of speech that are not merely "locutionary" but "illocutionary" or "perlocutionary." This is especially important for the centers of research and when science is related to other human activities (social, cultural, economic, ecological, political, etc.).

Language has a role in science *ad intra* but also *ad extra*. In this regard, as the crisis caused by the spread of the Covid-19 virus has highlighted,

---

[48] That we do things with words is something that initially discussed by several analytical philosophers: Austin (1962); Strawson (1970); and Searle (1969).

It is interesting that John L. Austin translated into English (with reproduction of the German text) a very important Frege book: Frege (1884). It is also worth remembering the volume that Strawson edited related to thought and action: Strawson (1968).

science does not belong only to scientists, the organizations that develop it and the entities — public or private — that finance it. In communicating science, in addition to doing locutionary acts (emission of statements with sense and reference), we also do illocutionary acts that have effects (e.g., establish actual or estimated facts regarding a treatment or a vaccine to lessen or stop the spread of the pandemic). Moreover, we do perlocutionary acts to try to persuade, dissuade or convince the general public, research-related organizations, or governments of the appropriate actions within given contexts, either spatial (region, country, international level, …) or temporal (days, weeks or months).

These speech acts, in their *ad extra* facet, can have incidence, effects and consequences of very diverse types: professional, legal, social, economic, cultural, political, etc. In this regard, there can be a negative impact when communicating science, if the speech content of the locutionary act is false, if the illocutionary acts are insincere, or if the perlocutionary acts are wrong. This kind of speech acts affects how the public perceives the research made from how it is communicated. Scientific activity is connected with other human activities, but it is known through the language used. Therefore, it has to aspire to correspondence with reality, truthfulness in the transmission of contents, and be a useful guide for other human actions.

Undoubtedly, we can follow the suggestion in philosophy of science in order to adopt "a change of focus from propositions to actions" (Chang 2014, 67). But scientific actions are already mediated linguistically: we select ends, means and potential outcomes based on the language we have to select them. Meanwhile, in scientific activity we are interested in what actions we take with words, such as illocutionary acts (e.g., committing ourselves to a research, keeping confidentiality if requested by the sponsor, or not publishing the research without the express consent of the publisher who publishes the work). Also perlocutionary acts, by which we can persuade to avoid a line of research or to follow an alternative option, to convince someone to join the research group or to dissuade the population with scientific data from irresponsible actions, when there is a highly traceable epidemic in the contagion.

When scientific language is used properly, it is distinguished from ordinary language. Thus, it aspires (a) to univocity about the sense expressed or, at least, absence of ambiguity; (b) to offer precision regarding the details reflected in its content; and (c) to have accuracy with reality concerning the reference, so that its ideal is exactness in terms and statements. This leads to a distinction between approaches such as scientific realism and what is ordinarily understood as "common sense" (cf. Sankey 2020, 68–83). Furthermore, scientific language is constantly used in technology, but the specific technological language is different, because is oriented towards the creative transformation of the real, resulting in a product or artifact that has a tangible and, as such, a reality that is identifiable and reidentifiable in a clear way.[49]

7 Origin and Structure of the Present Book

Some of the papers of this book have their origin in the Conference on Language and Scientific Research: The Role of Language in Basic and Applied Science (*Jornadas sobre Lenguaje e investigación científica: Papel del lenguaje en la Ciencia Básica y la Ciencia Aplicada*), held at the University of A Coruña, Ferrol Campus, on 9 and 10 March 2017. The main speaker was Jeffrey Barrett of the University of California at Irvine. The papers selected of this conference have been adapted for the purposes of this volume. In addition, several other papers have been incorporated in order to cover the field discussed.[50]

---

[49] On the relations between science and technology, with their consequent philosophical-methodological differences, see Gonzalez (2005).

[50] This book is added to the books coming from previous congresses, which are grouped in the *Gallaecia Series: Studies in Contemporary Philosophy and Methodology of Science*: *Progreso científico e innovación tecnológica*, 1997; *El Pensamiento de L. Laudan. Relaciones entre Historia de la Ciencia y Filosofía de la Ciencia* 1998; *Ciencia y valores éticos*, 1999; *Problemas filosóficos y metodológicos de la Economía en la Sociedad tecnológica actual*, 2000; *La Filosofía de Imre Lakatos: Evaluación de sus propuestas*, 2001; *Diversidad de la explicación científica*, 2002; *Análisis de Thomas Kuhn: Las revoluciones científicas*, 2004; *Karl Popper: Revisión de su legado*, 2004; *Science, Technology and Society: A Philosophical Perspective*, 2005; *Evolutionism: Present Approaches*, 2008; *Evolucionismo: Darwin y enfoques actuales*, 2009; *New Methodological Perspectives on Observation and Experimentation in Science*, 2010; *Conceptual Revolutions: From Cognitive Science to Medicine*, 2011; *Scientific Realism and Democratic Society: The Philosophy of Philip Kitcher*, 2011; *Las Ciencias de la Complejidad: Vertiente dinámica de las Ciencias de Diseño y sobriedad de factores*, 2012, *Creativity, Innovation, and Complexity in Science*, 2013; *Bas van Fraassen's Approach to Representation and Models in Science*, 2014; *New Perspectives on Technology, Values, and Ethics: Theoretical and Practical*, 2015; *The Limits*

A clear purpose to the book is to highlight the relationships between language and scientific research, so it seeks to fill a gap in the specialized literature. This explains the summary of the volume, which has an opening chapter that provides the historical and theoretical context of the volume and continues through four parts: (I) The Problem of Reference and Potentialities of the Language in Science; (II) Language and Change in Scientific Research: Evolution and Historicity; (III) Scientific Language in the Context of Truth and Fiction; and (IV) Language in Mathematics and in Empirical Sciences. Each of these parts has two chapters.

Within a thematic scheme that begins in general to reach a more delimited terrain, the book opens with this chapter on "The Relevance of Language for Scientific Research," where some key points of the relationship between language and scientific research are dealt with. They are (i) language as one of the constitutive elements of science; (ii) the main two roles of language in the configuration of science; (iii) the differences shaped by the language between basic science, applied science and application of science; (iv) the need for dynamic components for scientific language, because it cannot be reduced to structural components; (v) the dimension of language in scientific activity and the role of language of this activity connected with other human activities in social life.

Part I begins with "Semantics of Science and Theory of Reference: An Analysis of the Role of Language in Basic Science and Applied Science," by Wenceslao J. Gonzalez. First it goes to the core of the semantics of science, which is the way to deal with scientific language, and follows two different routes: the genuinely semantic and characteristically pragmatic. They are often based on conceptions originated in philosophy of language, which are adapted to the philosophy of science. Secondly, it addresses a key issue for the relationship between language and scientific research, which is the problem of reference. In this respect, the two main lines of solution are considered: (a) the position associated with semantic role and semantic value, and (b) the pragmatic approach, where the reference is the use made by someone of a term in a context. These approaches

of Science: An Analysis from "Barriers" to "Confines", 2016; *Artificial Intelligence and Contemporary Society: The Role of Information*, 2017; *Philosophy of Psychology: Causality and Psychological Subject. New Reflections on James Woodward's Contribution*, 2018; and *Methodological Prospects for Scientific Research: From Pragmatism to Pluralism*, 2020.

have an impact on the characterization of basic and applied science, so the chapter considers whether the two orientations mentioned have points in common and, if not, what fits better with what scientific language *is* and *ought to be* for scientific research.

How language affects scientific research is analyzed in the paper by Ladislav Kvasz: "On the Role of Language in Scientific Research: Language as *Analytic, Expressive, and Explanatory* tool." He tries to give an overview of the potentialities of language of science: (a) *analytic power*, which shows how complex formulas the language allows us to derive, formulas which could not be derived in the previous stages; (b) *expressive power*, which shows what new things the language allows us to express, things which in the previous stages of development defied expression; (c) *methodical power*, which shows what new methods the language enables us to introduce there, where in the previous stages of development we saw only several unrelated tricks; (d) *integrative power*, which shows what sort of unity and order the language enables us to see there, where in the previous stages we perceived just unrelated particular cases; (e) *explanatory power*, which shows how the language allows us to explain the failures which occurred in the previous stages, failures that were previously incomprehensible; and (f) *constitutive power*, which shows how the language, by transgressing the rules of its own syntax, enables us to constitute some radically new kinds of objects. All illustrations are taken from physics, but he thinks this is valid for a wide range of scientific disciplines.

Thereafter, part II starts with "Scientific Inquiry and the Evolution of Language" by Jeffrey Barrett. He holds that inquiry does not consist in choosing a fixed descriptive language, specifying a set of hypotheses in that language, then updating one's degrees of belief in the various hypotheses under consideration as one gets on new empirical evidence. Rather, our beliefs about the physical world and the language we use to express these beliefs coevolve. In this regard, Barrett considers how one might model the coevolution of predictive theory and descriptive language using the tools of evolutionary game theory. The result is a belief-revision model of knowledge that illustrates how subsequently evolved languages might exhibit semantic drift, invention, and discard.

"Language, History and the Making of Accurate Observations" by Anastasios Brenner also emphasizes the dynamic dimension of scientific language. He maintains that there is no neutral language of observation, nor a clear-cut dichotomy between observation and theory. Thus, some sort of consensus has been reached to the effect that facts are theory-laden, but the exact meaning of this assertion remains a controversial issue. He analyzes the bold claim by Lorraine Daston and Elizabeth Lunebeck that "scientific observation lacks its own history." The aim of Brenner's paper is to examine both the contributions and shortcomings of the new trends with respect to our understanding of observation in the sciences. In particular, it appears important not to dissociate scientific practice from the epistemic values such as accuracy that infuse it.

With part III comes "The Evolution of Truth and Belief" by Jeffrey Barrett. His paper concerns how endogenous epistemic norms may themselves coevolve with one's language and theoretical commitments. To this end, Barrett models the coevolution of language and epistemic commitments using a hierarchical evolutionary game, a sort of generalized signaling game. He begins by considering how a notion of truth might coevolve with an associated descriptive language. When successful, the notion of truth one ends up with will have evolved to play a concrete role in the game. He then considers how agents might develop a language that allows them to characterize their beliefs and degrees of belief. Finally, he discusses evolutionary conditions under which a principle of indifference that assigns equal prior probability to each possibility that one can describe in one's evolved language being in fact be successful.

Tarja Knuuttila in "Models, fictions and artifacts" contrasts the fictional and artifactual approaches to modeling. She argues in favor treating models as artifacts, purposefully created entities. These models are constrained in view of answering a pending scientific question, motivated by theoretical or empirical considerations. In this regard, the artifactual approach focuses on how models are constructed by making use of culturally established representational tools, in their various modes and media. In this approach models have both abstract and concrete dimensions and it tackles the vague ontological status of fictional entities. Thus, models are not divided into nonconcrete (abstract or fictional) models and concrete models. The artifactual account does not separate model systems and their actual representations from each other.

Then part IV offers "On Mathematical Language: Characteristics, Semiosis and Indispensability" by Jesus Alcolea. The aim of this contribution is to analyze some of the characteristics of mathematical language, the role of semiotic modes in reproducing the effectiveness of mathematics in science, and the relation of the problem of indispensability to the one of the reasonable effectiveness. He maintains that, unlike natural languages, mathematical language is a rigorously defined and unambiguous one. This characteristic constitutes its greatest advantage: its complete lack of ambiguity. In addition, he emphasizes that the range of topics communicated in natural languages and those communicated in mathematical language differ in significant ways. Feelings and emotions are not expressed in mathematical terms, and imprecisely defined terms are not allowed in mathematical language.

"Characterization of Scientific Prediction from Language: An Analysis of Nicholas Rescher's Proposal" by Amanda Guillan focuses on this pragmatic conception, which gives primacy to the view of meaning as use. In this regard, scientific prediction is then the result of an activity that seeks to obtain justified answers to meaningful questions about future occurrences. Within this framework of the primacy of pragmatics, the paper offers an analysis of the predictive statements in Rescher's proposal. Thus, 1) the attention goes to his proposal about prediction as a statement, where the features of the predictive statements are studied and the timing feature is analyzed, so the problem of retrodiction is also considered. 2) The existence of different types of scientific prediction, which Rescher does not develop in an explicit way. In this regard, we can see differences between the language of basic science and the language of applied science. 3) His approach on the limits of prediction and language is analyzed, both regarding the barriers or frontiers between the scientific predictive language and the non-scientific predictive statements and the confines or ceiling of the predictive language of science.

Finally, I would like to acknowledge the support of the people and institutions that have collaborated in the event that has served as the basis for the texts oriented to this book. Among the authorities, I would like to highlight the Rector of the University of A Coruña and the Vice-Rector of the Campus of Ferrol and Social Responsibility, who in 2017 supported the Congress and the creation of the Center for Research in Philosophy of Science and Technology, which is located in the Faculty of Humanities and Information Science.

In addition, I would like to recognize the support of several entities for the *XXII Conference on Contemporary Philosophy and Methodology of Science*: the Diputación of A Coruña, the City Hall of Ferrol, the Santander Bank, and the Foundation Philosophy of Science and Technology. My appreciation includes the Society of Logic, Methodology, and Philosophy of Science in Spain for its academic endorsement. I cordially thank the work done by Jessica Rey, Amanda Guillan and Pablo L. Vara for the edition of this book.

References

Andersen, H., Barker, P., & Chen, X. (2006). *The cognitive structure of scientific revolutions*. Cambridge: Cambridge University Press.

Austin, J. L. (1962). *How to do things with words*, edited by J. O. Urmson and Marina Sbisà. Oxford: Clarendon Press.

Azzouni, J., & Bueno, O. (2016). True nominalism: Referring versus coding. *British Journal for the Philosophy of Science, 67*(3), 781–816.

Carnap, R. (1931). Die logizistische Grunlegung der Mathematik. *Erkenntnis, 2*, 91–121.

Chang, H. (2014). Epistemic activities and systems of practice: Units of analysis in philosophy of science after the practice turn. In L. Soler, S. Zwart, M. Lynch, & V. Israel-Jost (Eds.), *Science after the practice turn in the philosophy, history and social studies of science* (pp. 67–79). New York: Routledge.

Clark, D. D. (2018). *Designing an internet*. Cambridge, MA: The MIT Press.

de Regt, H. W. (2017). *Understanding scientific understanding*. Oxford: Oxford University Press.

Dummett, M. (1975). What is a theory of meaning? (I). In S. Guttenplam (Ed.), *Mind and language* (pp. 97–138). Oxford: Clarendon Press.

Dummett, M. (1976). What is a theory of meaning? (II). In G. Evans & J. McDowell (Eds.), *Truth and meaning* (pp. 67–137). Oxford: Clarendon Press.

Dummett, M. (1977). *Elements of intuitionism*. Oxford: Clarendon Press.

Dummett, M. ([1973] 1981). *Frege: Philosophy of language*. London: Duckworth, 2nd ed. (1st ed. 1973).

Dummett, M. (1981a). *The interpretation of Frege's philosophy*. London: Duckworth.

Dummett, M. (1981b). Frege and Wittgenstein. In I. Block (Ed.), *Perspectives on the philosophy of Wittgenstein* (pp. 31–42). Oxford: Blackwell.

Dummett, M. (1991). *Frege: Philosophy of mathematics*. London: Duckworth.

Dummett, M. (1993). *Origins of analytical philosophy* (reprinted 1998). Cambridge, MA: Harvard University Press.

Floridi, L. (2014). *The Fourth revolution - How the infosphere is reshaping human reality*. Oxford: Oxford University Press.

Frege, G. (1884). *Die Grundlagen der Arithmetik. Eine logischmathematische Untersuchung über den Begriff der Zahl*. Breslau: Koebner. Translated into English by J. L. Austin (1950). *The foundations of arithmetic* (with reproduction of the German text). Oxford: B. Blackwell, (reprinted in 1978).

Frege, G. (1892). Über Sinn und Bedeutung. *Zeitschrift für Philosophie und philosophische Kritik, 100*, 25–50. Reprinted in G. Frege (1967). *Kleine Schriften*, (pp. 143–162), edited by I. Angelelli. Darmstadt: Wissenschaftliche Buchgesellschaft.

Frege, G. (1918). Der Gedanke. *Breiträge zur Philosophie des deutschen Idealismus, 1*, 58–77. Reprinted in G. Frege (1967). *Kleine Schriften* (pp. 342–362) edited by I. Angelelli. Darmstadt: Wissenschaftliche Buchgesellschaft.

Gadamer, H. G. (1960). *Wahrheit and Methode*. Tübingen: J. C. B. Mohr.

Gadamer, H. G. (1975). Hermeneutics and social science. *Cultural Hermeneutics, 2*, 307–316.

Gillies, D. A. (2012). The use of mathematics in physics and economics: A comparison. In D. Dieks, W. J. Gonzalez, S. Hartmann, M. Stöltzner, & M. Weber (Eds.), *Probabilities, laws, and structures* (pp. 351–362). Dordrecht: Springer.

Gonzalez, W. J. (1986a). *La Teoría de la Referencia. Strawson y la Filosofía Analítica*. Salamanca-Murcia: Ediciones Universidad de Salamanca y Publicaciones de la Universidad de Murcia.

Gonzalez, W. J. (1986b). El problema de la referencia en la Filosofía Analítica. Estudio bibliográfico. *Thémata, 3*, 169–213.

Gonzalez, W. J. (1991). Mathematics as activity. *Daimon, 3*, 113–130.

Gonzalez, W. J. (1996). Towards a new framework for revolutions in science. *Studies in History and Philosophy of Science, 27*(4), 607–625.

Gonzalez, W. J. (1998a). Prediction and prescription in economics: A philosophical and methodological approach. *Theoria: An International Journal for Theory, History and Foundations of Science, 13*(32), 321–345.
Gonzalez, W. J. (1998b). P. F. Strawson's moderate empiricism: The philosophical basis of his approach in theory of knowledge. In L. E. Hahn (Ed.), *The philosophy of P. F. Strawson* (pp. 329–358). Open Court, La Salle: The Library of Living Philosophers.
Gonzalez, W. J. (1999). Ciencia y valores éticos: De la posibilidad de la Ética de la Ciencia al problema de la valoración ética de la Ciencia Básica. *Arbor, 162*(638), 139–171.
Gonzalez, W. J. (2001). Lakatos's approach on prediction and novel facts. *Theoria: An International Journal for Theory, History and Foundations of Science, 16*(42), 499–518.
Gonzalez, W. J. (2002). Caracterización de la "explicación científica" y tipos de explicaciones científicas. In W. J. Gonzalez (Ed.), *Diversidad de la explicación científica* (pp. 13–49). Barcelona: Ariel.
Gonzalez, W. J. (2003). El empirismo moderado en Filosofía Analítica: Una réplica a P. F. Strawson. In J. L. Falguera, A. J. T. Zilhão, C. Martínez, & J. M. Sagüillo (Eds.), *Palabras y pensamientos: Una mirada analítica / Palavras e Pensamentos: Uma perspectiva analítica* (pp. 207–237). Santiago de Compostela: Publicaciones Universidad de Santiago.
Gonzalez, W. J. (2004a). La evolución del pensamiento de Popper. In W. J. Gonzalez (Ed.), *Karl Popper: Revisión de su legado* (pp. 23–194). Madrid: Unión Editorial.
Gonzalez, W. J. (2004b). Las revoluciones científicas y la evolución de Thomas S. Kuhn. In W. J. Gonzalez (Ed.), *Análisis de Thomas Kuhn: Las revoluciones científicas* (pp. 15–103). Madrid: Trotta.
Gonzalez, W. J. (2005). The philosophical approach to science, technology and society. In W. J. Gonzalez (Ed.), *Science, technology and society: A philosophical perspective* (pp. 3–49). A Coruña: Netbiblo.
Gonzalez, W. J. (2006a). Novelty and Continuity in philosophy and methodology of science. In W. J. Gonzalez & J. Alcolea (Eds.), *Contemporary perspectives in philosophy and methodology of science* (pp. 1–28). A Coruña: Netbiblo.
Gonzalez, W. J. (2006b). Prediction as scientific test of economics. In W. J. Gonzalez & J. Alcolea (Eds.), *Contemporary perspectives in philosophy and methodology of science* (pp. 83–112). A Coruña: Netbiblo.
Gonzalez, W. J. (2008a). Economic values in the configuration of science. In E. Agazzi, J. Echeverria, & A. Gomez (Eds.), *Epistemology and the social.*

Poznan Studies in the Philosophy of the Sciences and the Humanities (pp. 85–112). Amsterdam: Rodopi.

Gonzalez, W. J. (2008b). Evolutionism from a contemporary viewpoint: The philosophical-methodological approach. In W. J. Gonzalez (Ed.), *Evolutionism: Present approaches* (pp. 3–59). A Coruña: Netbiblo.

Gonzalez, W. J. (2008c). Rationality and prediction in the sciences of the artificial: Economics as a design science. In M. C. Galavotti, R. Scazzieri, & P. Suppes (Eds.), *Reasoning, rationality, and probability* (pp. 165–186). Stanford: CSLI Publications.

Gonzalez, W. J. (2010). Recent approaches on observation and experimentation: A philosophical-methodological viewpoint. In W. J. Gonzalez (Ed.), *New methodological perspectives on observation and experimentation in science* (pp. 9–48). A Coruña: Netbiblo.

Gonzalez, W. J. (2011a). The problem of conceptual revolutions at the present stage. In W. J. Gonzalez (Ed.), *Conceptual revolutions: From cognitive science to medicine* (pp. 7–38). A Coruña: Netbiblo.

Gonzalez, W. J. (2011b). Conceptual changes and scientific diversity: The role of historicity. In W. J. Gonzalez (Ed.), *Conceptual revolutions: From cognitive science to medicine* (pp. 39–62). A Coruña: Netbiblo.

Gonzalez, W. J. (2012). Methodological universalism in science and its limits: Imperialism versus complexity. In K. Brzechczyn & K. Paprzycka (Eds.), *Thinking about provincialism in thinking* (Poznan Studies in the Philosophy of the Sciences and the Humanities, vol. 100) (pp. 155–175). Amsterdam and New York: Rodopi.

Gonzalez, W. J. (2013a). Value Ladenness and the value-free ideal in scientific research. In C. Lütge (Ed.), *Handbook of the philosophical foundations of business ethics* (pp. 1503–1521). Dordrecht: Springer.

Gonzalez, W. J. (2013b). The roles of scientific creativity and technological innovation in the context of complexity of science. In W. J. Gonzalez (Ed.), *Creativity, innovation, and complexity in science* (pp. 11–40). A Coruña: Netbiblo.

Gonzalez, W. J. (2013c). The sciences of design as sciences of complexity: The dynamic trait. In H. Andersen, D. Dieks, W. J. Gonzalez, T. Uebel, & G. Wheeler (Eds.), *New challenges to philosophy of science* (pp. 299–311). Dordrecht: Springer.

Gonzalez, W. J. (2015a). *Philosophico-methodological analysis of prediction and its role in economics*. Dordrecht: Springer.

Gonzalez, W. J. (2015b). On the role of values in the configuration of technology: From axiology to ethics. In W. J. Gonzalez (Ed.), *New perspectives on technology, values, and ethics: Theoretical and practical* (Boston Studies in the Philosophy and History of Science) (pp. 3–27). Dordrecht: Springer.

Gonzalez, W. J. (2020a). Novelty in scientific realism: New approaches to an ongoing debate. In W. J. Gonzalez (Ed.), *New approaches to scientific realism* (pp. 1–23). Boston and Berlin: De Gruyter. https://doi.org/10.1515/9783110664737-001.

Gonzalez, W. J. (2020b). Pragmatic realism and scientific prediction: The role of complexity. In W. J. Gonzalez (Ed.), *New approaches to scientific realism* (pp. 251–287). Boston and Berlin: De Gruyter. https://doi.org/10.1515/9783110664737-012.

Gonzalez, W. J. (2020c). Pragmatism and pluralism as methodological alternatives to monism, reductionism and universalism. In W. J. Gonzalez (Ed.), *Methodological prospects for scientific research: From pragmatism to pluralism*, Synthese Library (pp. 1–18). Cham: Springer.

Gonzalez, W. J. (2020d). Levels of reality, complexity, and approaches to scientific method. In W. J. Gonzalez (Ed.), *Methodological prospects for scientific research: From pragmatism to pluralism*, Synthese Library (pp. 21–51). Cham: Springer.

Gonzalez, W. J. (2021). Semantics of science and theory of reference: An analysis of the role of language in basic science and applied science. In W. J. Gonzalez (Ed.), *Language and scientific research* (pp. 41–92). Cham: Palgrave Macmillan.

Gonzalez, W. J., & Arrojo, M. J. (2019). Complexity in the sciences of the internet and its relation to communication sciences. *Empedocles: European Journal for the Philosophy of Communication, 10*(1), 15–33. https://doi.org/10.1386/ejpc.10.1.15_1.

Hacking, I. (1999). *The social construction of what?* Cambridge, MA: Harvard University Press.

Hall, W., Hendler, J., & Staab, S. (2016). *A manifesto for Web science @10*, 1–4. Retrieved May 16, 2018, from http://www.webscience.org/manifesto.

Hendler, J., & Golbeck, J. (2008). Metcalfe's law, web 2.0, and the semantic web. *Journal Web Semantics: Science, Services and Agents on the World Wide Web, 6*(1), 14–20.

Hendler, J., & Hall, W. (2016). Science of the world wide web. *Science, 354*(6313), 703–704.

Hendry, D. F. (2012). Mathematical models and economic forecasting. Some uses and mis-uses of mathematics in economics. In D. Dieks, W. J. Gonzalez,

S. Hartmann, M. Stöltzner, & M. Weber (Eds.), *Probabilities, laws, and structures* (pp. 319–335). Dordrecht: Springer.
Husserl, E. (1901 and 1902). *Logische Untersuchungen.* Max Niemeyer, Hall a.S.: Max Niemeyer, Erster Tail, 1901, und Hall a.S.: Max Niemeyer, Zweiter Tail, 1902.
Kuhn, Th. S. ([1962] 1970). *The structure of scientific revolutions.* Chicago: The University of Chicago Press.
Kuhn, Th. S. ([1983] 2000). Commensurability, comparability, communicability. In P. D. Asquith & Th. Nickles (Eds.), *PSA 1982. Proceedings of the 1982 biennial meeting of the Philosophy of Science Association* (pp. 669–688), vol. 2, Philosophy of Science Association. East Lansing, MI; reprinted in Th. S. Kuhn (2000), *The road since structure: Philosophical essays, 1970–1993, with an autobiographical interview* (pp. 33–53). Chicago: University of Chicago Press.
Lakatos, I. (1970). Falsification and the methodology of scientific research programmes. In I. Lakatos & A. Musgrave (Ed.), *Criticism and the growth of knowledge* (pp. 91–196). Cambridge: Cambridge University Press; reprinted in I. Lakatos (1978), *The methodology of scientific research programmes. Philosophical papers, vol. 1* (pp. 8–101). Cambridge: Cambridge University Press.
Melo-Martin, I., & Intemann, K. (2018). *The fight against doubt: How to bridge the gap between scientists and the public.* Oxford: Oxford University Press.
Morrison, M. (2015). *Reconstructing reality. Models, mathematics, and simulations.* New York: Oxford University Press.
Nagel, E. (1961). *The structure of science. Problems in the logic of scientific explanation.* New York: Harcourt, Brace and World.
Niiniluoto, I. (1993). The aim and structure of applied research. *Erkenntnis, 38*(1), 1–21.
Niiniluoto, I. (1995). Approximation in applied science. *Poznan Studies in the Philosophy of the Sciences and the Humanities, 42,* 127–139.
Niiniluoto, I. (2020). Interdisciplinarity from the perspective of critical scientific realism. In W. J. Gonzalez (Ed.), *New approaches to scientific realism* (pp. 231–250). Boston and Berlin: De Gruyter.
Popper, K. R. (1935). *Logik der Forschung.* Vienna: Julius Springer Verlag.
Popper, K. R. (1945a). *The open society and its enemies.* Vol. 1: *The spell of Plato.* London: George Routledge and Sons.
Popper, K. R. (1945b). *The open society and its enemies.* Vol. 2: *The high tide of prophecy: Hegel, Marx and the aftermath.* London: George Routledge and Sons.
Popper, K. R. (1957). *The poverty of historicism.* London: Routledge and Kegan.

Psillos, S. (2010). Scientific realism: Between Platonism and nominalism. *Philosophy of Science, 77*(5), 947–958.
Putnam, H. (1962). What theories are not. In E. Nagel, P. Suppes, & A. Tarski (Eds.), *Logic, methodology and philosophy of science* (pp. 240–251). Stanford: Stanford University Press.
Rescher, N. (1977). *Methodological pragmatism: A systems-theoretical approach to the theory of knowledge*. Oxford: Blackwell; New York: New York University Press.
Rescher, N. (1996). *Process metaphysics*. Albany, NY: State University New York Press.
Rescher, N. (1999). *Razón y valores en la Era científico-tecnológica*. Barcelona: Paidós.
Rescher, N. (2003). Collective responsibility. In N. Rescher (Ed.), *Sensible decisions. Issues of rational decision in personal choice and public policy* (pp. 125–138). Lanham, MD: Rowman and Littlefield.
Salmon, M. H. (1992). Philosophy of the social sciences. In M. H. Salmon et al. (Eds.), *Introduction to the philosophy of science* (pp. 404–425). Englewood Cliffs, NJ: Prentice Hall.
Salmon, W. C. (1990). *Four decades of scientific explanation*. Minneapolis: University of Minnesota Press.
Salmon, W. C. (1998). *Causality and explanation. N.* York: Oxford University Press.
Salmon, W. C. (2002a). Explicación causal frente a no causal. In W. J. Gonzalez (Ed.), *Diversidad de la explicación científica* (pp. 97–115). Barcelona: Ariel.
Salmon, W. C. (2002b). Estructura de la explicación causal. In W. J. Gonzalez (Ed.), *Diversidad de la explicación científica* (pp. 141–159). Barcelona: Ariel.
Sankey, H. (2020). Scientific realism and the conflict with common sense. In W. J. Gonzalez (Ed.), *New approaches to scientific realism* (pp. 68–83). Boston and Berlin: De Gruyter.
Searle, J. R. (1969). *Speech acts: An essay in the philosophy of language*. Cambridge: Cambridge University Press.
Sen, A. (1986). Prediction and economic theory. In J. Mason, P. Mathias, & J. H. Westcott (Eds.), *Predictability in science and society* (pp. 3–23). London: The Royal Society and The British Academy.
Simon, H. A. (1991). *Models of my life*. New York: Basic Books (reprinted in The MIT Press, Cambridge, MA, 1996).
Simon, H. A. (1996). *The sciences of the artificial* (3rd ed.). Cambridge, MA: The MIT Press, (1st ed., 1969; 2nd ed., 1981).

Simon, H. A. ([1990] 1997). Prediction and prescription in systems modeling. *Operations Research, 38*, 7–14; reprinted in H. A. Simon, *Models of bounded rationality. Vol. 3: Empirically grounded economic reason* (pp. 115–128). Cambridge, MA: The MIT Press.

Strawson, P. F. (1950). Truth (II). *Proceedings of the Aristotelian Society, 24*, 129-156.

Strawson, P. F. (1961). Perception and identification. *Proceeding of the Aristotelian Society, 35*, 81–120. Reprinted in P. F. Strawson (1974), *Freedom and resentment and other essays* (pp. 85–107). London: Methuen.

Strawson, P. F. (Ed.). (1968). *Studies in the philosophy of thought and action*. Oxford: Oxford University Press.

Strawson, P. F. (1970). Phrase et acte de parole. *Langages, 17*, 19–33.

Strawson, P. F. (1979). Perception and its objects. In G. F. Macdonald (Ed.), *Perception and identity* (pp. 41–60). London: Macmillan.

Strawson, P. F. (1998). Reply to Wenceslao J. Gonzalez. In L. E. Hahn (Ed.), *The philosophy of P. F. Strawson*. The Library of Living Philosophers (pp. 359–360). La Salle, IL: Open Court.

Suppe, F. (1974). The search for philosophic understanding of scientific theories. In F. Suppe (Ed.), *The structure of scientific theories* (pp. 1–241). Urbana, IL: University of Illinois Press, (2nd ed. 1977).

Suppes, P. (1981). The plurality of science. In P. Asquith & I. Hacking (eds.), *PSA 1978*, Philosophy of Science Association, vol. 2 (pp. 3–16). East Lansing, MI: Philosophy of Science Association. (It was reprinted in Suppes, P. (1984). *Probabilistic metaphysics*. Oxford: B. Blackwell, Oxford (reprint in 1985), pp. 118 134, and in Suppes, P. (1993). *Models and methods in the philosophy of science: Selected essays* (pp. 41–54). Dordrecht: Kluwer.)

Thagard, P. (1992). *Conceptual revolutions*. Princeton: Princeton University Press.

Thagard, P. (2009). The cognitive structure of scientific revolutions. *British Journal for the Philosophy of Science, 60*(4), 843–847.

The Economist. (2020, May 9). High-speed science. The pandemic has caused scientists to work faster. That should be welcomed, p. 10. Section *Leaders*.

Tiropanis, T., Hall, W., Crowcroft, J., Contractor, N., & Tassiulas, L. (2015). Network science, Web science, and Internet science. *Communications of ACM, 58*(8), 76–82.

Toulmin, S. E. (1953): *The philosophy of science. An introduction*, London: Hutchinson University Library (3rd reprint, 1957).

Toulmin, S. E. (1971). From logical systems to conceptual populations. In R. C. Buck & R. S. Cohen (Eds.), *In memory of R. Carnap* (pp. 552–564). Dordrecht: Reidel.

Part I

The Problem of Reference and Potentialities of the Language in Science

2

Semantics of Science and Theory of Reference: An Analysis of the Role of Language in Basic Science and Applied Science

Wenceslao J. Gonzalez

Scientific research has in language a key factor, which takes many forms and is accessible to every researcher and scientific organization.[1] Semantics of science deals with this central component of scientific activity and analyzes the issues around meaning. Among them is the reference, which is especially important for the problem of truth. The theory of reference in the case of science considers, first, how and why scientific terms relate

[1] Language is a universal capacity that is expressed through many different codes, so language can never be intrinsically private. This is also the case of science, which is also a human capacity open to all and which cannot be purely individual. The horizon of the semantics of science moves in these coordinates, which advance on the basis of cooperation, just like science as a whole, where values play a role. Cf. Rolin (2015).

W. J. Gonzalez (✉)
Center for Research in Philosophy of Science and Technology,
University of A Coruña, Ferrol, Spain
e-mail: wenceslao.gonzalez@udc.es

W. J. Gonzalez (ed.), *Language and Scientific Research*,
https://doi.org/10.1007/978-3-030-60537-7_2

to objects or processes. This establishment of the reference as a relationship between language and reality (actual or possible) leads to a second problem: the contribution of the referent in the significance of scientific terms (and, consequently, of the statements of science). The third problem is how this reference relationship and the delimitation of the referent's contribution to meaning is transmitted so that the reference is incorporated by the scientific community.

These three problems of reference have to do with the role assigned to scientific terms and statements in basic and applied science. The options here are very varied. Initially, they depend on whether the main stream of the semantics of science follows a semantic line or a pragmatic path. Thereafter, they depend on how the role of reference in science is conceived: (i) as twofold with a semantic role and semantic value; (ii) as connected to an idea of causality in the use of terms related to objects; (iii) as part of a semantic anti-realist setting, where proof is the key notion instead of truth; and (iv) as an ingredient of a linguistic approach with taxonomies and semantic holism. Consequently, there is the impact of the previous steps on basic and applied science, on both meaning and reference.

1 Semantics of Science and the Theory of Reference

Within the philosophical-methodological study of science, semantics of science is one of the central parts. It deals with issues related to language, such as meaning, reference and truth,[2] considering that language is one of the constitutive elements of science.[3] Thus, semantics of science accompanies the other parts of this intellectual endeavor, such as logic of science, epistemology, methodology of science, ontology of science, axiology

[2] Issues of theory of meaning related to science configure only a part of the questions that are philosophically relevant to scientific language, as can be seen in this book. See, in this regard, Gonzalez (2021).

[3] These constitutive elements are the basis on which the topics are then discussed, which are in relation to issues such as objectivity, autonomy, critical attitude or progress of science. On these four aspects in contemporary science, see Niiniluoto (1984, 4–7).

of research, and ethics of science. These philosophical-methodological spheres deal respectively with the topics related to structure, knowledge, processes (procedures and methods), activity, ends and values. In addition, the issues raised by the semantics of science are also of interest in addressing questions of technology[4] and certainly have implications for the social dimension of science.[5]

1.1 Semantic Problems

From a thematic viewpoint, the semantic problems of science come with scientific language, either formal or natural,[6] and can be addressed at various levels: (i) science in general, insofar as they affect what a multivariate range of scientific disciplines have in common; (ii) groups of sciences, such as the empirical sciences investigating nature, society or the artificial; and (iii) particular sciences, in any of these three directions (physics, sociology or the sciences of the Internet [Tiropanis et al. 2015]), both in what they share with other disciplines (e.g., in terms of interdisciplinarity) and in what is distinctive or unique to that particular science. Thus, questions related to the language of scientific prediction can be raised with respect to science in general, groups of sciences (natural, social or artificial) or a particular science.[7]

Issues of the semantics of science can be raised according to a thematic scale, from the most abstract to the most concrete level. Thus, this scale begins with linguistic questions associated with macro-theoretical frameworks, such as those that concerned Thomas Kuhn (1962 [1970]) or Imre Lakatos ([1970] 1978), it then passes through questions related to scientific theories (basic or applied) and various types of models

[4] The relationships between these parts of the philosophy and methodology of science and the philosophy of technology are addressed in Gonzalez (2013a).

[5] This social dimension of science is addressed by the external perspective of the philosophy of science. Its link with the issues raised by technology is clear. Cf. Gonzalez (2005).

[6] For several decades, the emphasis on the semantics of science has been associated with approaches related to formal language, since obviously formal language needs interpretation that gives it a semantic content. The perspective of analysis here is much broader than that view.

[7] These three thematic levels of science — general, group and particular — in terms of scientific prediction are expressly addressed in the book Gonzalez (2015).

(descriptive and prescriptive), to arrive at hypotheses and contrasting processes via observation and experimentation. This wide range of options includes language issues in science being associated with certain thematic contexts, which have an impact on the interpretation of the content.[8] A representative case is "evolution," which can have a good number of contexts (general, a group of sciences, a particular science), as well as a set of possible interpretations (such as Charles Darwin, Stephen Jay Gould, etc.)[9] with respect to diverse reality levels (macro, meso and micro).

Semantic analysis can focus on the three main directions of scientific activity: basic, applied and of application. (1) Basic science is when semantic content is connected to the task of expanding human knowledge through problem-solving, in the search for truth or, at least, truthlikeness. (2) Applied science has a semantic content related to solving concrete problems within a practical setting. (3) Application of science is when the attention goes to the use of semantic content according to diverse contexts where scientific knowledge can be pertinent.[10] The type of issues varies from the first to the third case. The semantic line has had much more influence in the case of basic science, while a more pragmatic path is accepted in the case of applied science and is reinforced in the application of science (where a characterization of semantic content related to speech acts is also possible).

1.2 Reference: Relationship, Referent, and Transmission

Overall, the theory of reference deals with the relations between language and reality (actual or possible). First, it can focus on the relationship as

---

[8] The relevance of context is highlighted in philosophical-methodological trends such as pragmatism, which can approach the issue from "inside" to "outside," emphasizing the *intentionality* in the use of language, or from "outside" to "inside," which insists on the role of the *environment* around of the language used. A complete semantic analysis of the issue requires both the "inside" and the "outside" perspectives.

[9] On the history of the concept of "evolution" and its philosophical impact, see Gonzalez (2008a). See also Bowler ([1983] 2009).

[10] These three possibilities, in the case of scientific prediction, are considered in Gonzalez (2015, 32–40).

such, that is, how and why the links between terms and statements are with respect to objects or processes (past, present or future). Second, it can consider the contribution of reality itself, that is, the referent in its various forms of presentation that are grasped by scientific language. In this regard, the structural or dynamic properties of the real can be situated at the various levels of the real, such as the macro, meso and micro levels. Third, it can analyze how that reference is conveyed in the language of scientific activity, so that it can be part of a community of researchers that transmits a specific terminology used in scientific statements.

Originally, the theory of reference was developed in the philosophy of language, but it has had a broad thematic spectrum from the beginning. Hence, it has at least semantic, logical, cognitive and ontological components, as can be seen in analytical philosophy.[11] How and why words — expressions and statements — are linked to objects and processes is at the heart of the reference problem, which is where the relationship with reality matters, and connects to the issue of the semantic contribution of the reality itself. In this regard, the reference problems can be related to questions about meaning, the logical structure of the proposition, human knowledge and extramental reality (both past and existing as well as possible future). Each of them can have an impact on the semantics of science, the logic of science, the epistemology and the ontology of science.

Before going into the field of science, the theory of reference is a philosophical field where at least realism and idealism, Platonism and psychologism, conceptualism and nominalism compete. The two extremes are the reference as a pure intralinguistic relationship (the idealistic vision where the linguistic sense configures the referred object or process) and the reference as a mere extralinguistic element (the nominalism understood as that which highlights the other than language — the object or

[11] Those four thematic spheres are in Peter F. Strawson. A comparison between his conception and the positions of Gottlob Frege, Bertrand Russell, Ludwig Wittgenstein and Willard van Orman Quine is developed in Gonzalez (1986a).

The publications of these five philosophers and the works published about them are collected in Gonzalez (1986b). Within analytical philosophy, these publications address the theory of reference in relation to the theory of meaning, logical-linguistic aspects, the theory of knowledge and metaphysics.

the process — since, between the term used and the referent, there is no genuine mediation of the sense or a conceptual component).

Historically, the way in which the problem of reference acquires prominence in science is through the formal sciences and Frege's reflections on language. Specialists in Frege's logic and philosophy of logic, such as Peter Geach and Max Black, chose to translate the Fregean term "Bedeutung" as *reference* (Geach and Black [1952] 1960; see also Geach 1987). In this regard, Michael Dummett points out that *Bedeutung* "is simply the German word for 'meaning': but one cannot render 'Bedeutung' as it occurs in Frege, by 'meaning', without a very special warning. The word 'reference' does not, I think, belie Frege's intention, though it gives it much more explicit expression: its principal disadvantage is that it has also become customary to translate the cognate verb 'bedeuten' by the non-cognate verbal phrase 'stand for'. This tradition is unfortunate, but it is established" (Dummett 1981a, 84).

1.3 Meaning and Reference

Concerning the characterization of meaning in Frege, which is central to understanding its relationship to the reference, Dummett distinguishes de facto two approaches in this philosopher. The first is what he calls "the intuitive notion of meaning" (Dummett 1981a, 83), whereas the second is formed by the "elements in the meaning of a sentence or expression" (Dummett 1981a, 2). I mention this difference between both, because Dummett attributes to the first three ingredients (among them, force), while in the second there are only two elements (sense and tone).

On the one hand, Dummett considers that

> Frege drew, within the intuitive notion of meaning, a distinction between three ingredients: sense, tone and force. That is to say, he distinguished between these three things. He does not use any word to express the notion of 'meaning', as I have here used the word, and therefore does not claim sense, tone and force as being ingredients in anything more general. (…) Therefore, we may reasonably say that Frege discerns three ingredients within the intuitive notion of meaning: or, perhaps, better, that he proposes

> to replace the intuitive notion of meaning by the three notions of sense, tone and force. (Dummett 1981a, 83–84)

But, on the other hand, Dummett writes that

> Frege distinguished two elements in the meaning of a sentence or expression, for one of which he reserved the word 'sense' ('*Sinn*'), and for the other of which we might use the word 'tone' ('illumination' ['*Beleuctung*'] and 'colouring' ['*Färbung*'] being the words Frege himself used for this latter). He explained the difference in this way: to the sense of a sentence belongs only that which is relevant to determining its truth or falsity; any feature of its meaning which cannot affect its truth or falsity belongs to its tone. Likewise, to the sense of an expression belongs only that which may be relevant to the truth or falsity of a sentence in which it might occur; any element of its meaning not so relevant is part of its tone. (Dummett 1981a, 2)

Therefore, the difference between the two characterizations of meaning factors in Frege is in the idea of "force." It can be, for example, assertive or interrogative, which points to the idea of a use of language. In turn, it seems clear that the reference is not an "ingredient" or "element" of meaning in Frege.[12] In addition, "reference" is not the same as "stand for" and, within his threefold notion of "meaning" (sense, tone and force), reference is especially related to the sense, since — in his approach — only *Sinn* is relevant to truth or falsity. Thus, a word expresses (*ausdrücken*) a sense (*Sinn*) and refers (*bedeuten, bezeichnen*) a reference (*Bedeutung*). Furthermore, the reference is seen mainly as I) a relation with a "semantic role" and as II) a referent with "semantic value" (cf. Dummett 1978c, 120).

Together with the establishment of the reference, which entails picking out of something in order to single out it (that is, identify it), and assuming that there is, in principle, something that serves as a basis for making the reference, we arrive at the problem of *transmission* of the reference. Ian Hacking puts it this way: "Frege thought that a word has a standard

[12] "Reference, for Frege, is a notion required in the theory of meaning — in the general account of how language functions — just as the notion of truth is required: but the reference of a term is no more part of what is ordinarily understood as its meaning than the truth-value of a sentence is" (Dummett 1981a, 84).

sense, which is what makes a scientific tradition possible. The sense is what is shared by all communicators, and may be passed down from generation to generation of students" (Hacking 1983, 76).

Hacking exemplifies this with a museum visitor who sees a largish, preposterously shaped skeleton and says '*that* is what I mean.' For Frege, that skeleton would be the reference for a statement like 'the glyptodon brought by Richard Owen from Buenos Aires.' The *sense* of "glyptodon" would be a huge animal, which became extinct in South America, similar to the armadillo, with a peculiar set of teeth. As the statement makes sense, it enables us to pick out the *reference*, if there is any. Knowing the definition of "glyptodon" — the sense — I can go to the museum and look for its skeleton — the reference — if there is one, without having to look at the comments that may be next to the pieces on display.[13]

De facto, the problem of reference plays an important role in the contemporary debates of currents of philosophy of science regarding empirical sciences and scientific realism (Gonzalez 2020a). This is the case of structural realism, with the "ontic" version (Ladyman 2007, 2011). But it is also addressed in other approaches to scientific realism, such as the *divide et impera* stratagem or selective realism (Peters 2014), the critical realism defended by Ilkka Niniluoto (1999), the referential realism considered by Stathis Psillos (1999), the entity realism defended by Ian Hacking (1999),[14] or in pragmatic realism, as I conceive it (Gonzalez 2020b). But the topic of the reference has also interested authors of many other tendencies, such as the anti-realist one of naturalist orientation (cf. Laudan 1984) or the structuralist conception of scientific theories (cf. Balzer 1987; Moulines 1982).

[13] Cf. Hacking's chapter "Reference" in his book *Representing and Intervening*, especially, 75–76.
[14] See also Hacking, I., "Reference," in Hacking (1983 [2005], 75–91).

2 Two Main Directions: Semantic Line and Pragmatic Path

When addressing the problem of the role of language in science, the focus of the approach to the semantics of science and the theory of reference can follow two main directions: (i) the semantic line and (ii) the pragmatic path, which can intertwine depending on the characteristics of each being more or less accentuated. De facto, they lead to different interpretations of the role of language in basic science, in applied science and in application of science. The first line insists on the primacy of *semantic content* over context of use, so to be able to master meaning is, in principle, to be able to grasp the conditions when it is true. In contrast, the second path emphasizes *meaning as use*, so — instead of the conditions of truth — what matters are the conditions of proper use of the statements in a given context.

In the case of natural language, when dealing with the issues of meaning and truth, Peter F. Strawson characterized as a "Homeric struggle" the choice between the semantic line and the pragmatic path (Strawson [1970] 1971b). Originally, this distinction comes from the contemporary philosophy as a contrast that, according to Strawson, occurs between two approaches: (1) the theorists of formal semantics, who think that the essential is the content and the contingent is the use, where he situates Gottlob Frege and Ludwig Wittgenstein's early period, associated with the *Tractatus Logico-Philosophicus* (Wittgenstein 1922), where he places the linguist Noam Chomsky as well;[15] and (2) the theorists of communication-intention, thinkers that direct their attention towards intentionality in communication, such as John L. Austin, H. Paul Grice or Ludwig Wittgenstein in his late period (cf. Strawson [1970] 1971b, 172).[16]

[15] Chomsky connects his universal grammar with semantic criteria, cf. Chomsky (1972). He also suggested that pragmatic competence could be a cognitive system different from grammatical competence and endowed with a different structure, cf. Chomsky (1980). Chomsky has dealt with other topics on language that have unquestionable philosophical relevance, cf. Chomsky (1993), where he discusses his vision with other experts, and Chomsky (1995).

[16] He recognizes that the holders of one view "share some ground with their opponents" (Strawson [1970] 1971b, 176).

2.1 The Semantic Line

To follow a "semantic line" within the semantics of science involves the primacy of the *content* itself expressed through words and sentences. But it could also be through models and theories, insofar as they are understood as language or when they require an interpretation of content, because these are conceptions of a formal status when characterizing scientific models or theories. In fact, many of the approaches in favor of this semantic line are of a formal nature, either with a direct logical basis or proposing rational reconstructions of scientific theories with other abstract approaches. This kind of perspective, when there is a primacy of the content over the use of language, often analyzes the issues of sense and reference without giving special relevance to the context.[17]

Although this "semantic line" approach can be linked to various philosophical proposals over the years, Frege's distinction between sense (*Sinn*) and reference (*Bedeutung*) is particularly prominent in this line (Frege [1892] 1967a). This view has been influential in a large group of philosophical options regarding science, in general, a group of sciences, or some particular sciences. This influence can be noticed from the enunciative conception of scientific theories of neo-positivist inspiration to recent approaches of scientific realism that emphasize the role of reference.

But, in this long trajectory, these conceptions do not form a homogeneous block. Thus, within this general framework of preference for semantic content over the prevalence of usage, several conceptions are usually encompassed (Suppe [1974] 1977, 221–230).[18] First, the trajectory that began with logical neopositivism, passed through logical empiricism and became the "received view." Second, analyses proposed by Patrick Suppes on models can also be included in this group of semantic approach. Third, his analyses on models influenced on the

[17] An alternative version is the approach from the concept of "denotation," instead of "reference." This is a clearly different notion, which is why Peter Geach discarded it in order to translate *Bedeutung* in the case of Frege's approach. Cf. Geach, P. T. ([1958–59] 1970, 209, note).

[18] In this important study and in other publications, Bas van Fraassen's initial approach is situated within the semantic line. But there are more recent publications where he clearly incorporates the element of the pragmatic path, cf. van Fraassen (2008). On this conception, see Gonzalez (2014).

structuralist conception of Joseph Sneed, who was followed by Wolgang Stegmüller and later by Wolfgang Balzer and Ulises Moulines.[19] This conception poses an abstract core (K) and a set of intentional applications (I), which was first thought of for physics and has since been used to analyze and reconstruct scientific theories from other disciplines. But this list can be extended with a fourth option, consisting of approaches expressly based on probability theory.

(i) Over four decades, the trajectory followed from the logical neopositivism of the Vienna Circle to the decline of the received view, the semantics of science occupied a key place in connection with verification, verifiability and empirical confirmation (cf. Suppe [1974] 1977). "Verification" had a clear dual role: semantic and methodological. On the one hand, this led to an emphasis on empirical truth and, additionally, the confusion between meaning and truth; and, on the other, scientific progress was seen in linear and cumulative terms, which was later completely ruled out in the "historical turn," because it did not fit with the history of science. "Verifiability" allowed predictive statements to be meaningful and more in tune with the use of probability in science (cf. Reichenbach 1938). "Empirical confirmation" was another step forward for a better understanding of the links between theoretical language and observation language, within the framework of a structure of science that was beginning to open up (cf. Nagel 1961).

Most of the criticisms of this important semantic tradition came from issues on meaning and truth, sense and reference. Additionally, there are questions that concern another subject of interest to Frege: representation. In this regard, it can be pointed out that

> there is no doubt that the positivist analysis of theories is beset with serious problems and that certain non-linguistic elements such as structures do play an important role in scientific representation; but from this it does not follow that language *per se* is irrelevant to an analysis of scientific theories

[19] According to Moulines, Frege's semantic scheme is still the best tool for analysis at our disposal. Cf. Moulines (1982, 331). However, when he studies Frege's position in favor of truth as the main goal of science, Moulines is critical and in favor of the thesis of methodological instrumentalism. See Moulines (1990).

> or models. Scientific representation involves an intricate mixture of linguistic and non-linguistic elements and what we have come to understand is what this mixture is like and how the different parts integrate. (Frigg 2006, 62)

(ii) Suppes had his own version of what a semantic approach to science should be. So, at least since the early 1960s (Suppes 1962, 1967), he "has argued that the semantic approach of model theory is likely to be more fruitful than axiomatic approaches in analyzing theories, and has made some progress in determining the general features of such models" (Suppe [1974] 1977, 222). Regarding the semantic line, Suppes recognizes its merits:

> The great successes of set-theoretical semantics in mathematical logic, and the strong tradition, at least since Frege, against psychologism in logic, have been the basis for arguments for a similar approach to the syntax and semantics of natural language. If appropriate limits are clearly recognized, much can be and has been accomplished by set-theoretical methods. Some of the notable successes are the restricted use of Tarski's definition of truth and the semantic analysis of modal and temporal concepts. (Suppes "Language," in Suppes 1984 [1985], 149–150)

Nonetheless, Suppes also considers that "as a final account of meaning or reference, however, set-theoretical semantics in its standard form is clearly inadequate" (Suppes "Language," in Suppes 1984 [1985], 150). Thus, he seems to be clearly open to the pragmatic path:

> My general point is that we use sentences for many purposes. There is not just one literal use of sentences that requires a theory of fixed reference and definite truth. Rather, there are uses that are not straightforward and that require considerable circumspection to interpret. Moreover, as I have already emphasized repeatedly, it is hopeless to strive for an analysis that catches all nuances of spoken utterances. (Suppes "Language," in Suppes 1984 [1985], 150)

(iii) In 1971 Joseph Sneed developed some ideas that, according to Frederick Suppe, were "closely related" to Suppes' approach (Suppe

[1974] 1977, 223, note 558), and led to the structuralist conception of scientific theories (Sneed 1971). This perspective had an undoubted impact on the number of publications.[20] It was understood to be a semantic conception based on the idea of model inspired by Alfred Tarski and the model theory. For Sneed, scientific theory was articulated in core and applications. Within the nucleus, which was the key to characterizing the structure of a specific theory, a number of aspects were distinguished. Years later, Sneed — together with Balzer and Moulines — offered a more elaborate version in the book *An Architectonics for Science* (Balzer et al. 1987).

Among the features of the structuralist conception is the proposal of a formal apparatus that combines logic and set theory to elucidate the structure of all science, both from the synchronic and the diachronic angle (Balzer et al. 1987, p. XII). They focus on the formal structure of an empirical theory and develop a wide range of distinctions: potential models (Mp), actual models (M), partial potential models (Mpp), global constraints (GC) and global links (Gl) along with intended applications (I). The most elementary units in the architectonics of science are the "theory elements," where there is a certain "vocabulary" as a conceptual structure and some empirical law that uses this vocabulary, as well as the specifications to apply this law (Balzer et al. 1987, p. XX). This vocabulary is theoretical or non-theoretical (instead of observational, since it depends on the corresponding scientific theory) and is inserted within a network, in addition to being linked to other scientific theories.[21]

(iv) Other approaches with a formal component, which are based on probability theory and have had a great impact for a number of years, such as the Bayesian approach to probability, are focused on logical, epistemological and methodological aspects. Thus, they usually pay little attention to the semantics of science and the theory of reference. However, as with other formal approaches to analyzing scientific reasoning, the views regarding semantic content are commonly closer to the semantic line than to the pragmatic path. This seems to be Richard Jeffrey's point

[20] The publications of the first decades are collected in Diederich et al. (1989).

[21] On how this structuralist conception characterizes the unity of science, see in Echeverria (1990).

of view, when he saw the Bayesian perspective as "verificationism of a mild form" (Jeffrey 1975, 116; see also Jeffrey 1975, 113).

Later, when Colin Howson and Peter Urbach published *Scientific Reasoning: The Bayesian Approach*, their influential book on Bayesianism in personalist or subjective orientation, they left a testimony of their vision of the reference with a pragmatic component: "As probabilities are characteristically assigned to vernacular sentences and not to the formulas of formal languages, we shall accordingly use one and the same sentence for both specific and generic reference, distinguishing those uses by appropriate contextual stipulation" (Howson and Urbach 1989, 22).[22]

2.2 The Pragmatic Path

A full-fledged pragmatic path emphasizes the idea of meaning as use, where the language used is part of a practice and takes place within a context. Commonly, there are connections to a type of human activity and intentionality in communication. This pragmatic position on meaning admits, in turn, several possibilities, depending on the context of prevailing use. But the legacy of the late Wittgenstein — mainly his *Philosophische Untersuchungen* (Wittgenstein 1953 [1976, 2001]) — has had a clear impact on some philosophers of science, such as Stephen Toulmin. Wittgenstein's vision of language has some influence on Thomas Kuhn, who accepted the idea of meaning as use,[23] in addition to some repercussion on influential views of the external perspective on philosophy of science (in science, technology and society studies).

Altogether there are many approaches that, to a greater or lesser extent, accept a pragmatic position regarding the *meaning* of scientific terms and statements. Some of these approaches are worth noting: (1) positions

[22] *Scientific Reasoning: The Bayesian Approach* has a second edition in 1993 and a third one in 2006. See also Howson (2006). On this approach, see Gonzalez (2006, especially, 17–19).

[23] "Knowing what a word means is knowing how to use it for communication with other members of the language community within which it is current. But that ability does not imply that one knows something that attaches to the word by itself — its meaning, say, or its semantic markers. With occasional exceptions, words do not have meanings individually, but only through their associations with other words within a semantic field. If the use of an individual term changes, then the use of the terms associated with it normally changes as well" (Kuhn 1990, 301).

directly based on the intersubjective use of language, as in Toulmin's writings; (2) authors of the historical turn in philosophy and methodology of science where historical variability and the role of the scientific community are paramount, as is the case with Kuhn; (3) approaches to the "social turn" that are inspired by some Wittgensteinian and Kuhnian ideas to encourage a linguistic relativism in their publications on science, technology and society; (4) philosophers of science in tune with some of the lines of American pragmatism; and (5) thinkers in favor of a methodological pluralism who consider the support of a linguistic pluralism or, at least, discard a linguistic reductionism (either for science in general or for a group of sciences). Each of these proposals has its own approach to *reference* in terms of use of language: (i) mostly researches usage, (ii) predominantly historical, (iii) clearly social, (iv) specifically practical or (v) diversified according to many factors.

1) Undoubtedly, the most Wittgensteinian is Toulmin, especially when he wrote the book *Philosophy of Science* (Toulmin 1953 [1957]).[24] This harmony also led him to seek a better understanding of the cultural context in which the Viennese philosopher was formed (Janik and Toulmin 1973 [1996]). Following the analysis of the language of prediction, he goes deeper into this subject in one of the chapters of another book: *Foresight and Understanding* (Toulmin 1961, 18–43; see also Gonzalez 2010, 95–126). Toulmin held a prominent position at the 1965 Congress at Bedford College (London), where the Popper-Kuhn controversy took place. His presence there led him to a direct criticism of the Kuhnian positions of his initial stage, objections that he raised from a different philosophical option than the Popperian one (cf. Toulmin 1970).

Toulmin is Wittgensteinian in his understanding of language: meaning as use prevails, because "the adoption of a new theory involves *language-shift*" (Toulmin 1953 [1957], 13).[25] He is expressly interested in prediction to clarify the aims of science, which leads him to question the primacy of predictive success as the key task of science. Also, he is unequivocally critical of the Kuhnian distinction between "normal sci-

[24] Late Wittgenstein has an impact on several passages of the book, being one of the most cited authors: Toulmin (1953 [1957], 13–14, 51, 81, 88–89, 129, 162–163 and 172).

[25] In Kuhn the paradigm shift entails a change of meaning, which affects the question of incommensurability; cf. Kuhn (1962 [1970], 102).

ence" and "revolutionary science," which it does not see as genuine explanatory categories of scientific events (cf. Toulmin 1970, especially, 47).[26] But Toulmin considers "explanatory paradigms" in the 1961 publication (Toulmin 1961, 16, 43, 52–54, 56–58, 60–62, 66–67, 69–70, 74, 76–77, 81 and 103; see also Gonzalez 2010, 113–116), the year before Kuhn's famous book.

2) During the "historical turn," Kuhn is in tune with some Wittgensteinian ideas on the characterization of language.[27] The significance of scientific terms and statements in a pragmatic way is in his conceptual scheme on scientific change, where historicity plays a key role, whereas in the case in Wittgenstein's late period it has not any relevant position. Moreover, when Kuhn developed his three philosophical-methodological approaches (Gonzalez 2004), the first — *The Structure of Scientific Revolutions* — and the second — *Second Thoughts on Paradigms* (Kuhn [1974] 1977)) — give less relevance to language than the third (where a key text is *Commensurability, Comparability, Communicability* [Kuhn [1983] 2000a]).

Between Kuhn's first philosophical-methodological period of a historiographic type, in which he assumes de facto a linguistic relativism with his vision of the paradigm-shift in scientific revolutions, and in his third period, of a markedly linguistic character, there is a noticeable contrast. At this more mature stage, his view is close to the causal theory of the reference and to Putnam's internal realism. Thus, it is clearly distant from the initial radical incommensurability for an objective comparison of references. Now scientists can be "bilingual" and understand the change in scientific theories (Kuhn 1990).

---

[26] Cf. Toulmin (1970, especially, 47). A comparison of the two thinkers in terms of the characterization of the prediction is found in Gonzalez (2013b).

[27] Quite interesting is the recognition that Kuhn made in this third period on 'could one play chess without the queen?': "Twenty-five years ago the quotation was a standard part of what I now discover was a merely oral tradition. Though clearly 'Wittgensteinian,' it is not to be found in any of Wittgenstein's published writings. I preserve it here because of its recurrent role in my philosophical development and because I've found no published substitute that so clearly prohibits the response that the question might be answerable if only there were more information" (Kuhn 1990, 316–317, note 15).

3) Often under the influence of Kuhn's initial stage — which included the pragmatic path of language, epistemological relativism and the relevance of scientific communities — the "social turn" occurred, but later he criticized its excesses.[28] This trend include some philosophical views, such as certain versions of naturalism of sociological roots and most of the postmodern conceptions of science, which have emphasized the sociological dimension of scientific activity. In this regard, there have been three very influential positions within the sociology of science that follow the pragmatic path on scientific language (cf. Gonzalez 2006, 9–11):

(a) The "strong program" of the Edinburgh school led by Barry Barnes (1977, 1982, 1995; and Barnes et al. 1996) — later at the University of Exeter — and David Bloor (1973, 1974, [1976] 1991, 1983, 1997), based on some Kuhnian as well as Wittgensteinian ideas of his late period; (b) the empirical program of relativism (EPOR) developed by Harry Collins (1983, and Collins and Pinch 1993), which studies scientific controversies with an interpretative flexibility and analyzes their connections to the socio-cultural milieu; and (c) the ethnomethodology — the study of the actors' network at the workplace — defended by the social constructivism of Bruno Latour (1988, 1993; and Latour and Woolgar 1986), and Steve Woolgar (1981, 1988a, b).

4) Contemporary pragmatism has many faces, but methodological pragmatism has in Nicholas Rescher a pioneer and a key figure (Rescher 1977), who has also reflected on the origin of this important philosophical-methodological trend and its adaptation to the present time (Rescher 2012, 2014). His philosophical approach has insisted on the primacy of practice, so scientific progress is associated with "evaluating ways of doing things on the basis of their functional efficacy in realizing the collective objectives" (Rescher 2020, 69). Rescher's pragmatic idealism defends a *conceptualist idealism* and accepts a *pragmatic* vision of language (Rescher 1992). His view is more in tune with the conditions of assertiveness than with the conditions of truth as the key to grasping the meaning of statements and being able to identify the appropriate reference.

---

[28] "I am among those who have found the claims of the strong program absurd: an example of deconstruction gone mad" (Kuhn [1991b] 2000d, 110).

Rescher has focused on

> a theme that was central to Peirce's interests, namely linguistic communication. And in its dealings with this topic it emphasizes an approach — that of cognitive economics, of the "economy of research" — that was very near and dear to Peirce's heart. Moreover, the conclusion that it seeks to draw from these deliberations is also strikingly Peircean in spirit, by way of showing that attention to the pragmatic properties of discourse can contribute to the resolution of philosophical difficulties. The motivating idea of this essay is that various philosophical problems resolve themselves in a natural way once we take proper account of the practical principles that govern our communicative use of language. (Rescher 1998b, 3)

5) Methodological pluralism also usually comes with the pragmatic path for scientific language, once three types of reduction — of language, of subject matter and of method — are questioned. Although it was towards the end of the 1970s that the trend towards methodological pluralism began to emerge, it has been gradually taking shape since then and in different directions, especially since 2006, when the book *Scientific Pluralism* was published (Kellert et al. 2006). Actually, there are important reasons to criticize the methodological approaches to monism, reductionism and universalism and to move in favor of methodological pluralism:[29]

(i) The monist aspiration to a single and full account of reality is increasingly more difficult due to the development of new sciences (mainly, in the social and the artificial realms, such as the sciences of the Internet) [Gonzalez and Arrojo 2019]. (ii) Any attempt to reduce one science into another one can have certain epistemological, methodological and ontological costs.[30] (iii) The search for a universal method, valid in principle for all the sciences does not fit the variety of objects and problems that can be discussed, many of them completely new that lead to new methods instead of a priori methods or the mere convergence of

[29] A development of these reasons can be found in Gonzalez (2020c).

[30] On the issue of the "formal" and "nonformal" conditions for reductions in science, see the influential analysis made in Nagel (1961, 354–366). This double set of conditions can be used to consider the philosophico-methodological costs of the reduction.

the methods already available. (iv) There is an increasing difficulty of justifying a branch in any science that might be "fundamental" for the whole discipline. (v) Contextual factors of procedures and methods of scientific activity are relevant for methodology of science. In addition, there is a recognition of a plurality of stratagems for making research in empirical sciences (natural, social and of the artificial).

3 The Role of Reference in Science

Regarding the reference, it seems of particular interest to consider two relevant approaches of the semantic line and two others of the pragmatic path. The first option of the semantic line in science is associated with the *semantic role*, when the reference as a relation between the term and the object or process designated, and the *semantic value*, when the reference as a reality designated by the term used.[31] This position leads us to connect semantic, epistemological and ontological realms in basic science and in applied science, especially in some versions of the scientific realism. The second option is a realist alternative that came with the causal theory of reference,[32] which has had in Hilary Putnam an influential thinker as well as important philosophers of language. Later Putnam's positions shifted towards the more problematic approach of "internal realism."

Besides the influence of the later Wittgenstein on meaning and truth, the pragmatic path received an idea inspired by P. F. Strawson on referring against Bertrand Russell view on denoting (Strawson [1950] 1971a), which include that the reference is the use made by someone of a term in a context. In the pragmatic path, two appealing options are considered here. First, the anti-realist semantics, developed by Michael Dummett, when he utilizes the Wittgensteinian interpretation of language to rethink the intuitionist mathematics; and, second, Kuhn's linguistic approach of his third stage, which also considers Wittgensteinian ideas and became close to the internal realism.

[31] These two aspects are dealt with by Dummett (1981a, 210–211).

[32] "Causal insofar as it explains a person's use of a term with a certain reference in terms of a causal nexus between this use and earlier uses of that term with that reference" Kroon (1985, 143).

3.1 Reference as Twofold: Semantic Role and Semantic Value

Semantically, the role of reference in scientific expressions and statements is in principle twofold: (i) there is a semantic role, when the focus is on the relationship of reference between the terms and the objects or processes designated by those words, and (ii) there is a semantic value, when the reality itself (actual or potential), which is designated by the expression used, comes to the fore. In principle, it can be thought that both the semantic role and the semantic value have a role in the three main types of statements: explanatory, predictive and prescriptive.[33]

If the semantic line is followed, the *semantic role* consists of the properties to be ascribed to the constituent expressions of each given type, so that it can be determined that the statement made is true or not true. Thus, two different scientific expressions (e.g., two plants bearing the names of the different places they were found) have the same reference when they have the same importance for the truth in the statements in which they appear. This would be something like the potential truth value. In any case, the semantic role has to do with the interchangeability *salva veritate*: the reference of an expression E coincides with the reference of the expression E' if they are interchangeable while preserving the truth (e.g., two chemical elements discovered in parallel and initially named differently).

According to Dummett,

> to say that, in general, the semantic value of a singular term is an object does not in any way restrict the kind of semantic theory we adopt (unless some particular ontology, some doctrine about the kinds of object which the world contains, is presupposed); hence Frege's original argument does not depend upon the adoption of a classical, two-valued semantics. But the extension of the argument to other atomic sentences depends upon assuming that the semantic value of a predicate is its extension, i.e., its being determinately true or false of each object in the domain; and this holds good only within the classical semantics. (Dummett 1978c, 133)

[33] In the case of the application of science, we can add other types of statements of a more contextual type, as happens in medicine with preventive statements in individualized therapy.

Semantic value comes from that which is real (natural, social or artificial), which must be ascribed to the expressions that compose the statement in order to determine that statement as true or not true. Thus, the semantic value complements the semantic role, as it is something of an appropriate type that is associated with the semantic role. It happens that Frege, besides attributing reference to the subject of the statement, also attributes it to the predicate of the statement. When it is a singular term, the semantic value would be in the referred object; while, when it is a predicate, it would be its extension (that is, the objects of which it is true). In such a case, 'Covid-19 is a type of coronavirus' the semantic value is in both parts of the sentence — subject and predicate — but it is different.

Reference is articulated with the sense, which gathers the way in which the reference is given or presented as such. This allows us to capture different faces of reality, as proposed by methodological pluralism. Sense is then the way to access the reference. In addition, the sense — core of the meaning — is objective and has a cognitive character. Through the distinction and relations between *Sinn* and *Bedeutung* there is a version of semantic realism.[34] It is based on two particularly important theses: (1) meaning can be explained from the truth-conditions; and (2) reference occupies a relevant position understood as "semantic value," which accompanies the "semantic role" (Gonzalez 1986a, 106–109).

It is important to emphasize that, for Frege, *sense* has a cognitive character that makes possible the noetic or intellectual incorporation of the extramental real. At the same time, *Sinn* is *the way* to reference, so that it establishes a relationship between words and objects. In this respect, sense can be considered as a silent mediation. The semantic role and the semantic value of reference complete the picture. Through the role, the elements relevant to the truth (the properties), which are ascribed to the expressions, are taken into account. Through the value, what counts is the reality itself. Thus, the real features are considered and, associated to an expression, make a statement in which it appears to be true.

[34] In addition to Michael Dummett, with different nuances, other specialists adhere to Frege's realistic interpretation: Ignacio Angelelli, Gregory Currie, Richard Eldridge, Peter Geach and Michael Resnik, thus discarding the vision of Hans Sluga. See Gonzalez (1986a, 25–40; especially, 25–27).

All this — the combination of sense, semantic role and semantic value — rendered into scientific theories leads to the fact that these theories — and, therefore, their macro-theoretical frameworks — are commensurable. There is the possibility of having objective criteria to fix the content of scientific terms — and also of scientific statements — when two theories are in competition with respect to the same area of reality. Thus, when the macro-theoretical frameworks ("paradigms," "research programs," etc.) change while the terms of competing theories remain the same, it is always possible — according to the Fregean position — to establish the sense and reference of what is expressed, which allows an objective comparison between the old and the new theory.

Still the issue is complicated when it comes to scientific prediction, since it anticipates the future (ontological, epistemological or heuristic). If the focus is ontological, then the future is, by definition, that which it is not yet and, moreover, may become or never become a reality. In this regard, Dummett points out that the sense in Frege does not consist only in the way the object is given or presented, since there is a second ingredient of the sense: the "conventional meaning" (Dummett 1981a, 310–311 and 317–319). By means of the conventional meaning it can be understood that, even if the referent does not exist (the ether as a light-conducting medium or phlogiston as the key to combustion), the sense of a singular term ('the ether' or 'the phlogiston') can be understood by everyone who knows the language,[35] by those who are familiar with it. This possibility allows Frege to continue within an approach of semantic realism and avoid a Platonism (that is, to give entity to the sense itself, so that it always refers, although the object does not exist).

Although Frege gives objectivity to the sense, *Sinn* is not self-subsistent, it is not an end in itself, since it is conceived as a route for reference. He also warns that when the sense of an expression is captured, the reference is not always known (as happened with the gravitational waves predicted by Albert Einstein). Thus, having *Sinn* is not the same as having *Bedeutung*, as exemplified with the definite description 'the farthest celestial body from the Earth', which Frege assumed made sense, but doubted had any

[35] It should be noted that, for Dummett, practical knowledge and knowledge of language have relevant differences. Cf. Dummett, M. (1978d).

reference. The important thing here is that the consistency of sense is not *eo ipso* equivalent to the real existence of that around which one speaks or writes.

Here we can appreciate the difference between Frege's semantic line and Kuhn's pragmatic path, especially in his first philosophico-methodological stage. He assumes that what Frege calls "sense" and "reference" of scientific terms is de facto intrinsically dependent on the paradigms where it is found. Moreover, changing the paradigm means seeing a different world, even if the real things remain the same. It changes perception to the point where you can say that scientists respond to a different world after a scientific revolution.[36] The presence of the objective element in the sense of Frege avoids Kuhn's incommensurability, as he understood it in his stage of *The Structure of Scientific Theories*. Comparability is then possible, because the content of the sense of the scientific terms in question can be profiled and the reference relationship — and, where appropriate, the referent — can be established. This is diluted in the first Kuhnian stage, where in the end the revolutionary change associated with the paradigms — with one semantic holism replaced by another — is psycho-sociological (a "conversion").

Furthermore, if Frege were to admit that kind of Kuhnian semantic holism, where there is an intrinsic dependence on the accepted paradigm, *Sinn* would be reduced in principle to conventional meaning, with the consequent historicity of each scientific term used. But he does not follow the drift towards the semantic relativism. For Frege, what is relevant is the problem of *truth*, so that is the goal when making science. In this respect, the reference is more important than the sense, insofar as without reference true statements would not be possible. This is why he distinguishes between different linguistic uses: in order to do science, one must pay attention to the terms that have *reference*, while in other intellectual fields — such as poetry — expressions that have sense are valid, even if they have no reference.

[36] "In so far as their only recourse to that world is through what they see and do, we may want to say that after a revolution scientists are responding to a different world" (Kuhn 1962 [1970], 111).

3.2 Causal Theory of Reference and Internal Realism

Together with the influential Fregean semantic line, there are other versions of semantic realism, such as Hilary Putnam's conception based on the causal theory of reference. This is clear in his initial approach to realism, developed before 1976 (cf. Putnam 1973), and has special features in his stage of "internal realism" (cf. Putnam 1978b, 1983b). The causal theory of the reference also has prominent supporters, such as Saul Kripke or Keith Donnellan. In the case of the causal theory of reference, objectivity is appreciated in the referent, which has in principle an internal structure with characteristics that might be essential. Thus, the realistic character of this conception rests on the identity of the referent: it remains commonly constant from one occasion to another of reference, thus propitiating reidentification. The meaning, unlike Frege's approach, includes among its ingredients the reference, which comes to be the cause of the significance of the terms.

Although there is a continuity Putnam's semantic approach, insofar as he highlights the notion of "reference" as the axis for a realistic conception, he introduces important variations in its second stage. His initial idea was to emphasize the importance of the reference to determine the meaning, as opposed to the position — present, among others, in Frege — that insists on the meaning (especially, the "sense") as a determinant of the reference. Putnam then conceives the reference as dependent on causal connections. Language develops causal and non-causal links with different aspects of the world (cf. Gonzalez 1993, 35).

For Putnam, in his initial stage, there is a relationship between the sphere of language and thought, on the one hand, and reality, on the other, where the former corresponds asymptotically — at least to some extent — to the latter. In this respect, the theory of reference is then a theory of correspondence between these two poles (Putnam 1975, 290). Subsequently, and without considering his previous position wrong, Putnam has placed much greater emphasis on epistemological problems, obtaining consequences that he did not expect (cf. Putnam 1983a, VII–XVIII; especially, VII), which fall within the framework of "internal realism." The turn occurs when he notices that the notion according to which there is a "correspondence" between our words and determinable objects (objects that would have a determined reference independently of the conceptual scheme) is problematic, and he proposes another possibility, a situation that occurs in 1976.

At the beginning, the causal theory of reference was conceived as an alternative to approaches advocated by Bertrand Russell. In this regard, Putnam emphasizes that real things have a causal role in the *acquisition* and *use* of terms. It is accepted that a term refers to something if it remains in the correct relationship with those things existentially given (causal continuity in the case of proper names; identity of 'nature' in the case of class terms). In addition to the world's contribution to establishing the causal continuity in terms, there is a social cooperation: the expert must be consulted in order to know whether or not something is the indicated reference (e.g., gold), or it is necessary to resort to the chain of historical transmissions that persist in the reference of the term. Within these coordinates, one can avoid the problems of incommensurability of theories, which are so worrying in relativistic approaches (such as those proposed initially by Kuhn), since causal continuity presupposes the identity of the reference, so objective comparability of theories would be possible.

Along with his interest in the reference, Putnam has been noted for his constant attention to the truth. When he has turned to internal realism, he has continued to highlight these two notions as keys to understanding the relationship of language to the world, but has rejected the Fregean approach that equates understanding a statement with knowledge of its conditions of truth (cf. Putnam 1978b, 100). On the one hand, the reference has become "internal:" it depends directly on the theory in which it appears; and, on the other, truth has ceased to be understood in line with the idea of "correspondence," to be conceived as "idealized justification."[37] In both the first and, above all, in the second, Putnam's internal realism has received the impact of Dummett's anti-realism. The new trajectory is a third way between his previous realism and semantic anti-realism (cf. Gonzalez 1993, 36–37; see Putnam 1983b, 84).

Putnam's *internal realism* connects with a feature of semantic idealism to understand the notion of truth: "Idealists have always maintained that

[37] This view is criticized in Resnik (1987).

our notion of truth depends on our understanding of our theory and of the activity of 'discovering' it, *as a whole*. If I am right, then this is an insight of idealism that realists need to accept — though not in the way idealists meant it, of course" (Putnam 1975–76, 194).[38] This semantic holism of Putnam's internal realism connects with his alternative to the "external" type of metaphysical realism, which he condenses into three philosophical theses: a) the world consists of a fixed totality of objects independent of the mind; b) there is only one single and complete true description of the way the world is; and c) truth includes some sort of correspondence (cf. Putnam 1990, 30).[39]

That kind of metaphysical realism is of a classical nature, insofar as the primacy corresponds to the real and not to language. Following this framework, the world takes precedence over scientific theory. In this regard, internal realism emphasizes the dependence of reality on scientific theory. This implies a characterization of the reference — as a relationship between scientific terms and objects or processes — that is markedly intra-theoretical. Thus, this approach chooses to follow a very different direction to the "external" route: it is not possible to say how the world is independently of the scientific theory. Moreover, outside the scientific theory, the world becomes a mere "noumenon," in pure "thing in itself" (Putnam 1978a, 133).

Hence, internal realism rejects the three previous theses of metaphysical realism to sustain other positions. (i) The world is not something already elaborated but, in a certain way, made by the human being or processed through a conceptual scheme. Thus, the question about the components of the world only makes sense within a scientific theory or description. (ii) The world can be described in different ways, all of them might be true and complete even if they are rivals or opposites. (iii) Truth is not transcendent in its status but epistemic, so it depends on the theory (cf. Gonzalez 1993, 45).[40]

[38] This lecture took place on 23 February 1976, the year of the shift towards "internal realism," which was completed months later, in the lecture given on 29 December 1976: Putnam (1978a).

[39] This chapter has its origin in a 1982 text by Putnam, dedicated to two realists, which was revised for publication in the 1990 volume.

[40] In addition, Putnam addresses "Kant's problem, the problem of explaining the referential connection between our 'representations' and the world" where he criticizes Jerry Fodor's approach. See Putnam (1992, ch. 3, 35–59).

Many versions of scientific realism have been placed against this way of understanding realism,[41] with one of the most constant being Raimo Tuomela. Since 1979, he has criticized this approach, trying to show that the arguments given for rejecting that version of metaphysical realism and adhering to internal realism are far from convincing (cf. Tuomela 1979, 1985, 1987, 1990). On the one hand, Tuomela argues that, initially, Putnam is in a basically correct situation when he questions that version of metaphysical realism.[42] But, on the other, Tuomela considers that Putnam fails to defend his internal realism. Moreover, what he defends should be given another name, such as *coherentism* (or, even, perhaps *idealism*) [cf. Tuomela 1979, 124], because Putnam tends to consider reference and truth as intralinguistic notions: both depend strongly on the theory in which they appear.

Consequently, Tuomela considers that Putnam's internal realism abandons to a large extent what is proper to realism in order to participate in central features of idealism. He considers that Putnam's approach makes all semantic relations not only epistemic, but also, at the same time, strongly intratheoretic. Although Tuomela does not question that the elaboration of semantic relations has epistemic aspects, he thinks, however, that a realist should also accept that there are transtheoretic language-world relations. These relations might not be strictly semantic, but, in any case, they must be something related to this thematic area (cf. Tuomela 1979, 126). In this regard, Tuomela focuses his objections to Putnam on two frequent themes of scientific realism: reference and truth.

Based on his version of "critical scientific realism," Tuomela rejects the semantic elements of internal realism. He considers that, following the approach of internal realism, the statements about reference — which can then be related to the Tarskian conception of truth — become intratheorical *tautologies* for Putnam. Thus, a statement like '"cow" refers to cows' is — for him — a tautological metalinguistic statement (cf. Tuomela 1979, 126). Within this framework, Tuomela finds it striking that Putnam continues to defend that realism is the explanatory theory

[41] Although internal realism has left its mark, insofar as it has favored aspects such as pluralism in the analysis of science, it does not play a leading role in current trends in scientific realism. See, in this regard, Gonzalez (2006, 1–23).

[42] "I basically agree with Putnam's views on metaphysical realism" (Tuomela 1979, 124).

that gives reason to the events of science through the acceptance of reference as the characteristic of scientific terms and the admission of scientific theories as close to the truth.

Later, Tuomela insisted more on epistemological and ontological aspects, indicating that the Putnamian "metaphysical"-"internal" distinction is not original, since it is already available in Kant. He distinguished between "transcendental realism" and "empirical realism" (and the "transcendental idealism" associated with it). What Putnam calls *metaphysical realism* reflects Kantian transcendental realism; while *internal realism* would be the echo of Kant's empirical realism (cf. Tuomela 1990, 165). Tuomela also maintains that the Putnamian argument should not lead to the replacement of metaphysical realism in favor of internal realism but to a different situation. Thus, taking as a thread the interpretations of the concept of "truth," Tuomela discards the internal and metaphysical approaches. His position was then a "critical scientific realism," which was soon redirected towards a "causal internal realism" (cf. Gonzalez 1993, 46).

3.3 Semantic Anti-Realism and the Place of Reference

Even though Dummett has made an enormous contribution to realist semantics, developing multiple aspects of Frege's philosophy in three fields (language, logic and mathematics),[43] his personal philosophical conception in this realm is another approach: anti-realist semantics, which is related to central aspects of intuitionism (cf. Dummett 1977). It is based on a Wittgensteinian interpretation of the intuitionist mathematics of L. E. J. Brouwer ([1948] 1964, [1952] 1975a; see also Brouwer 1975b, 1981). This in turn raises the question of the very philosophy of mathematics of Wittgenstein's second stage, to see to what extent it is a conception close to Brouwer's mathematical intuitionism.[44]

---

[43] In addition to the above-mentioned publications, it is worth noting Dummett (1991).

[44] On this issue, see Gonzalez (1991a).

Conceptualism is the position that, in my judgment, underlies Dummett's anti-realist approach, since it gives priority to the idea of "construction" and recognition by the researcher as opposed to "discovery" and independence of the subject, in accordance with the inspiration of Brouwer's intuitionist mathematics. Thus, the emphasis is placed on "proof" — something constructed — as opposed to "truth," which is not the result of a construction but rather reached after a search, introducing an important change with respect to the realistic scheme: proof is temporary — and therefore revisable — while truth is not (Gonzalez 1990a).

Realism in Frege highlights that reality determines as true or false the statements (the propositions or "thoughts" they express), regardless of our knowledge or the ability to discover their truth value. When he explains meaning from the conditions of truth and relies on the notion of "reference" understood as a semantic value, reality has a preferential place. Its semantic realism highlights the link between sense (*Sinn*) and the condition by which the statement is true (the proposition or "thought" expressed) (cf. Frege [1918] 1967b).

Consequently, to master the *meaning* of a word — singular term or predicate — is to understand its contribution to fix the condition required for the statement containing it to be true, and the meaning expressed by the statement is the *sense* that the condition for it to be true has been fulfilled. It thus highlights the specifically semantic: the Fregean approach does not direct attention to what is known when the use of an expression is mastered, nor does it attend to the belief of the speaking subject; it wants to show something quite different: that it is from the internal structure of the statement that the proposition is determined as true. For Frege, the semantic values associated to the expressions are those that make possible the incorporation of reality and allow this semantic theory to be seen as objective.

Dummett emphasizes more the *pragmatic* component of language, which leads him to replace true statement with "justified assertion."[45] Basically, anti-realism is articulated in two theses contrary to realism: (i) it discards the explanation of meaning from the conditions of truth,

[45] Dummett's approach is not identical to that advocated by his follower Crispin Wright. Cf. Wright (1981, 1987).

which has to be replaced by a pragmatic vision of meaning where the axis is in the notion of "use;" and (ii) it devalues the importance of the reference, which is redirected to the role within the use of language instead of highlighting the semantic value contributed by the referent. The concept of "truth" is replaced by the idea of "proof," both for the formal sciences and for empirical sciences (cf. Dummett [1963] 1978a and 1978b; see also Dummett 1982).

Once this anti-realist semantics is considered in the case of the natural or social sciences, the proof then appears as a "justified assertion," a justification that, *sensu stricto*, is reduced to *empirical evidence*. Thus, Dummett's criticism of realism goes in two ways: (a) through substituting the notion of "truth," which is frequently considered as a key piece of semantic realism,[46] and (b) through rethinking the objective contribution of reality, which leads it to a position akin to verificationism (conceived as "verifiability" instead of "verification").

As regards the reference of scientific terms, Dummett's "anti-realism" is a far from idealistic approach. Both the origin of this semantic position and the role it assigns to the reference are inscribed within conceptions that are quite different from linguistic idealism, such as Brouwer's mathematical intuitionism and Wittgenstein's philosophy of language, according to which the mastery of a statement generally consists of the ability to use that statement. Following this position, the Fregean concept of "truth" — which attributes its objective possession or lack of it to each proposition that expresses a statement, regardless of our means to know such a value — cannot be central to the theory of meaning, and is replaced by the Wittgensteinian approach of meaning as use.

Thus, according to Brouwer and Wittgenstein, the mastery of the meaning of a mathematical statement lies in the ability to recognize, when dealing with each mathematical construction, whether or not it constitutes a proof of the statement, instead of consisting of the knowledge of what, independently of our means to know if it is so or not, must be given in order for the statement to be true (cf. Gonzalez 1990a,

[46] Scientific realism, in general, and semantic realism, in particular, have various expressions today. In this regard, objectivity is more important than truth, cf. Gonzalez (2020a).

150–152 and 163–165).[47] As the idea of *construction* and the *relevance of the use* of language prevail, the task of reference certainly has a more pragmatic and contextual character than the case of reference as a semantic value.

Here Dummett carries out two tasks through the Wittgensteinian interpretation of intuitionist mathematics: I) the constitution of anti-realist semantics, which possesses essentially different characters to those held in classical mathematics; and II) making Brouwer's positions — which minimizes the role of language, insofar as he conceives it as a mere instrument to transmit what is thought and to fix the mathematical constructions — compatible with the positions of Wittgenstein, who defends the relevance of language for the life and activities of man (among which is mathematics, which he sees from an anthropological perspective) (cf. Gonzalez 1991b).

To achieve both aspects, Dummett distances himself from Frege's central theses on meaning and conditions of truth. Thus, he opposes them with the primacy of *usage* and *justified assertion*, so the meaning and the problem of truth are understood in a pragmatic way. It also moves away, especially in mathematics, from the usual approach of realism as a study of entities, to consider it — and criticize it — from the perspective of language. This turn is due to Dummett's conviction of the pre-eminence of the theory of meaning over metaphysics (cf. Gonzalez 1990a, 153 and 165–166).

However, taking the reference back to linguistic practice — the mastery in the use of language — does not guarantee scientific objectivity. Moreover, as Colin McGinn has explained, there are cases in which, when use is explained as the exercise of recognition skills in relation to conditions of assertiveness, there are differences in content that are not accompanied by differences in use (McGinn 1982). It also happens that, when faced with a disjunction between 'p' or 'q', there are many cases in which we consider ourselves justified in affirming 'p or q', although we have no justification for asserting 'p' or for asserting 'q' (cf. Gonzalez 1990a, 160–161 and 169). An example is a bank counterfeit in the case

[47] On the choice between "truth" and "justifiability" and the shortcomings of the anti-realist approach in this respect, see Gonzalez (1990a, 159–162 and 169–170).

of two foreign currency notes, when we cannot assert whether the first or the second is counterfeit, while under true conditions one would be true and the other not.

3.4 Kuhn's Approach in his Linguistic Period

Kuhn, when he makes the "linguistic turn" after the second stage of *Second Thoughts on Paradigms*, makes an approach on semantics of science with important nuances regarding the previous positions. (1) He now considers that each scientific theory has its own lexicon, so it is a network that taxonomically orders the *concepts of genera* (kind-concepts) or the *terms of genera* (kind-terms), those that have reciprocal relations. (2) The scientific revolutions are now characterized as changes of meanings in the lexicons of the scientific theories, instead of being changes of paradigms (paradigm-shift). (3) The old distinction "normal science"-"revolutionary science" becomes the difference between activities that require changes in the lexicon and those that do not. The description of scientific revolutions in terms of changes in form (*Gestalt switches*) disappears. (4) "Incommensurability thus becomes a sort of untranslatability, localized to one or another area in which two lexical taxonomies differ" (Kuhn [1991a] 2000c, 93). This does not imply incomparability, since it is a local phenomenon. To understand a theory is not then to translate it into one's own language — in the manner of Willard van O. Quine — but to acquire a new language, which means becoming bilingual. (5) Different languages structure the world differently, so lexicons acquire the function of paradigms with respect to the world and experience (cf. Irzik 2001; Gonzalez 2004, 61–66).

Now, in the phase of the "linguistic turn," the scientific revolution does not make the reciprocal understanding of the contents of the alternative proposal impossible. There are concrete difficulties of translatability that are avoided by learning to be bilingual. For Kuhn, "members of one community can acquire the taxonomy employed by members of another, as the historian does in learning to understand old texts. But the process which permits understanding generates bilinguals, not translators" (Kuhn [1991a] 2000c, 93).

Reducing scientific revolutions to difficulties of translatability, within a "local" sort of incommensurability,[48] makes the relativistic vein — which was unmistakable in the texts of the initial period — diminish considerably. Incommunicability and incomparability are therefore ruled out: the revolution only entails difficulties in translation, so they do not have to be understood as "linguistic relativism." Moreover, in view of the relativistic excesses of the *strong program*, Kuhn warns at this stage of the need for the notion of "truth."[49] He does so, moreover, in the presidential address of the Philosophy of Science Association. There he expressly indicates that, "properly understood — something I've by no means always managed myself — incommensurability is far from being a threat to the rational evaluation of truth claims that it has frequently seemed" (Kuhn [1991a] 2000c, 91).

Three are now features of scientific revolutions, which is the frame for the new version of the historicity of science, in general, and of the reference, in particular. (i) Revolutionary changes are, in a way, holistic: they cannot be fragmentary (piecemeal), as is the case with normal or cumulative changes. Thus, in revolutionary change there are two options: either one lives with the incoherence or one reviews a number of interrelated generalizations (cf. Kuhn [1987] 2000b, 29). (ii) There is a change in language that "alters not only the criteria by which terms attach to nature but also, massively, the set of objects or situations to which those terms attach" (Kuhn [1987] 2000b, 29–30). (iii) Revolutions bring about changes in taxonomic categories. This change involves an adjustment of the relevant criteria for categorization but also of the way in which objects and situations distributed within the previous categories occur. Redistribution affects more than one category and, because they are interrelated, the change is holistic (it affects the whole set of categories) (cf. Kuhn [1987] 2000b, 30).

[48] The characterization of "local incommensurability" appears in Kuhn ([1983] 2000a, 35–37).

[49] "It is needed (…) to defend notions like truth and knowledge from, for example, the excesses of postmodernist movements like the strong program" (Kuhn [1991a] 2000c, 91).

Following scientific revolutions as categorical-linguistic, Kuhn describes himself as "post-Darwinian Kantian."[50] This qualifier is explained, on the one hand, by the similarity between the Kantian categories and the Kuhnian lexicon, since in both cases — categories and lexicon — they provide the preconditions for possible experience; and, on the other hand, the post-Darwinian feature is that Kuhnian lexicons — unlike the Kantian categories — are evolutionary. Moreover, lexicons have a historical and social nature: they change over time and as they move from one community to another.[51]

Looking from the point of view of the referent, it does not seem that Kuhn has achieved a coherent position, capable of overcoming the philosophical-methodological difficulties that a background duality carries: (a) the existence of a Kantian world in itself (*noumenon*), and (b) the Kuhnian thesis — explicit in its first stage — of the absence of a world of phenomena that has priority, since each one of them depends on a conceptual scheme (that can even *constitute* that world) (cf. Psillos 1994, 923–926; especially, 924–925). Moreover, consistency is made difficult by the fact that Kuhn sometimes accept and sometimes reject the existence of the Kantian "thing itself" (*Ding an sich*).[52]

One moment in which Kuhn distances himself from the Kantian position and moves towards a version of critical scientific realism is with the causal theory of reference, when he analyses Putnam's first stage. In this regard, Kuhn maintains that "excluding proper names, I doubt that there is any set of terms for which causal theory works properly; but it comes very close to doing so for terms like 'gold,' and the plausibility of its application to natural-kind terms depends on the existence of such terms.

[50] "The position I'm developing is a sort of post-Darwinian Kantianism. Like the Kantian categories, the lexicon supplies preconditions of possible experience. But lexical categories, unlike their Kantian forebears, can and do change, both with time and the passage from one community to another" (Kuhn [1991a] 2000c, 104).

[51] Evolutionism inspired by Ch. Darwin and historicity, understood in the strict sense, are not the same. On the notion of "historicity," cf. Gonzalez (2011).

[52] Acknowledging Kuhn that he defends the existence of Kantian *Ding an sich* in "The Road since *Structure*," the co-editors expressly point out: "Kuhn had earlier rejected the notion of a *Ding an sich* (see essay 8 ["Metaphor in Science," 1979]), and he again later repudiated (in conversations with us) both that notion and the reasons he had put forward for it" (Conant and Haugeland 2000, 7).

Terms that behave like 'gold' ordinarily refer to naturally occurring, widely distributed, functionally significant, and easily recognized substances. Such terms occur in the languages of most or all cultures, retain their original use over time, and refer throughout to the same sorts of samples. There is little problem about translating them, for they occupy closely equivalent positions in all lexicons. 'Gold' is among the closest approximation we have to an item in a neutral, mind-independent observation vocabulary" (Kuhn 1990, 309).

But we can raise the problem of reference in a plain case of identification, such as 'this is *a*,' which corresponds in principle to an ostensive definition of the kind and involves attending to the semantic route from "bottom up." In a way, this is the alternative route to semantic holism, which is Kuhn's favorite in its three stages. In his third period he still maintains that "I have been suggesting (…) that with occasional exceptions terms do not *individually* have meanings at all" (Kuhn 1990, 317, note 22). Meanwhile, "reference is a function of the share structure of the lexicon, but not of the varied feature spaces within which individuals represent that structure" (Kuhn 1990, 317, note 22). Moreover, he expressly admits to being close to Putnam's internal realism.[53]

Interestingly, "recognition statements" — as Dummett (1981a, 232) calls them — are basic to learning science: 'this is hydrochloric acid,' 'this is a comet,' 'this is a black hole.' Here the sense is needed to identify the reference indicated by "this," which includes the criteria of *identity* and *application*. By means of the criterion of identity we have the characteristics of the object or process, which allow us to know that it is the same or of the same type as another already known and points out the conditions that make the recognition statement true. The application criterion determines when it is correct to say 'this is a,' so it corresponds to the domain of knowing how to use the expressions.

Certainly, the criterion of identity is more important, since it allows us to determine when statements of the form 'this is the same comet that …' are true. The object or process that appears as referent cannot be recognized outside of language, because only through the domain of

[53] "Putnam has (…) moving (…) to a view ("internal realism") with significant parallels to my own" (Kuhn 1990, 317, note 23).

sense — which incorporates the criteria of identity and application — can we speak of concrete objects or processes. To avoid the problem of tautology, present in internal realism, we must admit the semantic value of the referent. When 'a' and 'b' are names or defined descriptions of the same object, insofar as they deal with the same reality (where it is true to say that 'a = b'), there is a difference in sense based on the different ways in which the object or process occurs. Thus, two identity statements such as 'water is H_20' and 'water is water,' while both being true, in the first case adds information, whereas the second is a tautology.

Accordingly, there is a primacy of the scientific language over the pure factuality of the phenomena in order to be able to identify the real and to have the recognition statements. This is true even if it is admitted that scientific language is based on the real (actual or possible). Moreover, the relations between sense and reference are very close. Because in establishing what an expression represents, we *say* what its reference is (e.g., neutrino), but not what its sense is. Instead, by choosing a particular way of doing this, we *show* what the sense is (e.g., subatomic particle), for the sense is a partial knowledge of the reference.[54] Scientific identification and recognition of what is known is guaranteed if the sense is objective and cognitive in its status, accompanied by criteria of identity and application.[55]

4 From Basic Science to Applied Science

Debates on the semantics of science and the theory of reference are usually focused on basic science (mainly physics and biology). These analyses consider explanatory or predictive statements rather than purely "descriptive" statements. De facto, they deal with two major possibilities: (a)

[54] This issue has been analyzed by Dummett among others. See Dummett (1981b).

[55] The next step — if we analyze it in terms of the configuration of theories with an ontological component — leads to how to approach the identity of these objects or processes — and therefore also their diversity — in two dimensions: the structural and the dynamic. Concerning the first, several philosophical-methodological viewpoints come into play, among which is ontological structural realism. See, in this respect, Ladyman (2007, especially, 23). Regarding the second — identity and diversity from a dynamic perspective — see Gonzalez (2013c).

statements in an environment of giving reasons for why something happens, where the referent is designated by the subject of the statement and the predicate attribute is sought to match the properties of the referent; or (b) statements in a situation of anticipating a possible future (ontological, epistemological or heuristic), the correctness of which is sought to be contrasted.

Commonly, in both cases the relationships between the semantics of science and the theory of reference include two relevant aspects: (i) a relationship between meaning and truth where an idea of correspondence is assumed for the statement considered true[56] or, at least, a notion of truth that might be compatible with the notion of correspondence,[57] and (ii) the real existence or potential of the referent designated by the scientific statement, depending on the explanatory or predictive statement. The second case opens the door to prescriptive statements, which is central in applied science. This is especially important in the sciences of the artificial that, like economics, are design sciences (cf. Gonzalez 2008b).

4.1 Semantic Differences

Philosophico-methodological debates on the language of applied science are much less frequent than in basic science. Thus, questions of the semantics of science and the theory of reference take on a more pragmatic character, since applied science is a question of solving concrete problems (cf. Niiniluoto 1993, 1995). The contextual aspect becomes more important, in order to make predictive statements that serve as a guide for subsequent action and prescriptive statements that provide guidelines for action, in accordance with certain values (cf. Gonzalez 1998, 2015,

[56] On the conceptions of truth, see Schmitt (2004). This book includes a final text: "Truth: A Bibliography," prepared by Kevin Kimble and Frederick F. Schmitt, pages 307–317. See also the critical analysis in Kirkham (1992).

[57] This is what happens with the notion of truth as redundancy, where — according to Peter Strawson — the idea of correspondence is *presupposed* when dealing with redundancy: "One who makes a statement or assertion makes a true statement if and only if things are as, in making that statement, he states to be" (Strawson [1970] 1971b), 180). Meanwhile, truth as redundancy is also present — in my judgment — in the vision of mathematics in the late period of Wittgenstein. Cf. Gonzalez (1987).

317–341). The concept of truth connects with practice in this sphere, indicating the right path to reach a desired goal. Here the notion of truth of pragmatism takes on more prominence (cf. Kitcher 2011), where what counts is the effective result that endorses the path followed as the appropriate one, under given circumstances.

Historically, there has been much influence of the semantic line in basic science, which has been traditionally associated with the use of *logic* as an instrument of analysis. In contrast, in the case of applied science, the semantic line has had a rather scarce presence, especially to the extent that applied science is associated with the idea of *construction* (and, in this respect, it moves away from the usual theses of scientific realism). In this regard, the semantic line has been more influential in the natural sciences than in the social sciences or the sciences of the artificial. This is particularly important in the case of the sciences of design, which are related to how things *ought to be* and are focused on *synthesis* rather than on analysis (Simon 1996, 4–5).

Looking back to philosophico-methodological analysis available, it seems clear that "it is by no means clear that conceptions founded upon the model of basic sciences would do justice to the peculiar characteristics of the applied sciences" (Niiniluoto 1993, 2). In this regard, approaches based on a pure semantic line cannot address properly the problems of applied science. The language of solving concrete problems of a practical or operational character — through prediction and prescription — requires clearly pragmatic factors, where the relationship with the context of use and the environment in which it is situated (social, institutional, economic, etc.) has an important weight.

Design sciences such as economics, information science, communication sciences or the sciences of the Internet constantly provide examples of the importance of pragmatic factors in scientific language, which is in line with the increasing complexity in the dynamic side (cf. Gonzalez and Arrojo 2019). In turn, the interaction between scientific creativity in the field of artificial sciences and the role of technological innovation, all within a social environment (cf. Gonzalez 2005), leads to the creation of new terms, the change of meaning of existing ones, etc. This novelty of semantic content, which can be longitudinal or horizontal — the extension of what is available — or transversal or vertical (something radically new), is compatible with linguistic objectivity for doing science.

Objectivity of semantic content implies that the content does not depend on the individual mind of the researcher, on the group of researchers in which he or she forms a scientific community, or on the institutions that support that research. It also implies that the semantic content has a series of features that allow its use by researchers all over the world and in prestigious accredited publications. Usually, basic science has greater recognition in this respect, which allows a term such as "phlogiston" to be discarded in favor of "oxygen." In applied science, this objectivity is achieved through practical ways of solving specific problems, as is the case in pharmacology.

4.2 The Perspective of Reference

Semantic value has a central task in basic science, insofar as it deals with what is real in the present or has been real in the past, whereas semantic role is reinforced in applied science, especially when there is a *range of possibilities* to solve concrete problems at stake. In this regard, a reference in mere terms of "internal realism" does not solve the problem of scientific progress rather it dissolves the problem, because reality, through the semantic value of the reference, has to be present in an ordinary way in order to do scientific research if we want to achieve genuine scientific progress (cf. Gonzalez 1990b, 2015, 29–32). Moreover, it cannot be the case that, in basic science, there is only intelligibility on the part of the scientific theory, and no intelligibility on the side of the reality.[58]

Applied sciences are related to epistemic and practical utilities (Niiniluoto 1993, 5). The specific purpose may be imminent or be in the short, medium or long term, or even very long term (as is the case with some climate change proposals). Within this practical domain, the reference as a semantic role and as a semantic value belong to science as a human activity, which is in turn connected to other human activities, dealing with concrete problems. Thus, the reference as relationship to a

[58] It seems odd to claim that "the epithet 'intelligible' applies to theories, not to phenomena" (de Regt 2017, 12). The same idea can be found in pages 45 and 88.

human undertaking and the reference as a referent in a dynamic scenario have to do with current or possible activities. A purely synchronic approach to the reference is not valid then, since the normal thing in applied science is that it has to be diachronic and with effectiveness in solving practical issues (in the natural, social or artificial world).

Diversity is a feature in basic science as well as in applied science. Now it is clear there is a need for methodological pluralism, which assumes the existence of a polyhedral reality (cf. Gonzalez 2020b). This means that the reference as a semantic role can designate aspects of the real from different angles, as is currently the case with the research of Covid-19. It also implies that the reference as a semantic value has to be a support point to have scientific objectivity, so that it allows true statements to be reached later on (for example, to cure an illness). These statements can be, in principle, to adequately reflect what is real (like the existence of numerous exoplanets) or to capture the correct path to achieve the concrete goals sought (as happened with penicillin and some diseases).

Historicity is another feature of both basic and applied science. But there is room for the identification of the referent — the synchronic aspect — and the subsequent reidentification of the referent (the diachronic component) as the same object or process previously identified. It is then possible (a) to establish the reference of a class of objects or processes and (b) that an adequate transmission of the reference can take place, so that the scientist or the research center can use the term to refer to the same type of reality even when there are temporal discontinuities.

Nevertheless, this feature is compatible with the incompleteness of the reference,[59] so that the designation of objects or processes may be incomplete, because reality is, in principle, richer and more varied than our language. Along with the mentioned polyhedral character of the real thing, there are anomalies, some aspects not predictable at the moment or have characteristics difficult to identify on a given occasion. This is one of the reasons why theories are commonly reformulated, new models are proposed or hypotheses are adapted, so that the reformulated semantic content allows a better reference relationship in the new statements.

[59] On the issue of the incompleteness of the reference, see De Groot (1987, 632–633 note).

To solve the problem of identification, the reference class should be established.[60] It is understood that it must not be empty (that is, that there are no fictitious entities such as "spontaneous generation") and preferably homogeneous, in that it is sought that this reference class shares elements of some kind. That there is a reference as a semantic value is required for it to be true, but it can certainly make sense even when it is not yet known whether there is a reference (as happened with the ontological prediction of the Higgs boson). This is why in applied science, insofar as it uses prescriptive models to propose a possible future to which we could or should go, it relies above all on reference as a semantic role. Because, when a prediction is stated it can be correct, but it can only be true once the event has taken place (cf. Rescher 1998a), which implies the presence of the reference as a semantic value.

Acknowledgment This paper has been developed within the framework of the project FFI2016-79728-P, supported by the Spanish Ministry of Economics and Competitiveness (AEI).

References

Balzer, W. (1987). Reference and development of theories. In P. Weingartner & G. Schurz (Eds.), *Logic, philosophy of science and epistemology* (pp. 213–224). Vienna: Hölder-Pichler-Tempsky.

Balzer, W., Moulines, C., & Sneed, J. (1987). *An architectonics for science*. Dordrecht: Reidel.

Barnes, B. (1977). *Interests and the growth of knowledge*. London: Routledge and K. Paul.

Barnes, B. (1982). *T. S. Kuhn and social science*. London: Macmillan (New York: Columbia University Press, 1982).

Barnes, B. (1995). *The elements of social theory*. Princeton: Princeton University Press.

Barnes, B., Bloor, D., & Henry, J. (1996). *Scientific knowledge. A sociological analysis*. Chicago: The University of Chicago Press.

[60] See, in this regard, Strawson's comments on when the objective reference emerge. They are made in his criticism to Quine's views. Cf. Strawson (1986, 519–532; especially, 521–531).

Bloor, D. (1973). Wittgenstein and Mannheim on the sociology of mathematics. *Studies in History and Philosophy of Science, 4*(2), 173–191.
Bloor, D. (1974). Popper's mystification of objective knowledge. *Science Studies, 4*, 65–76.
Bloor, D. (1983). *Wittgenstein: A social theory of knowledge*. London: Macmillan.
Bloor, D. ([1976] 1991). *Knowledge and social imagery*. London: Routledge and K. Paul (2nd ed., Chicago: The University of Chicago Press, 1991).
Bloor, D. (1997). *Wittgenstein, rules and institutions*. London: Routledge.
Bowler, P. J. ([1983] 2009). *Evolution: The history of an idea*. Berkeley: University of California Press; 25th Anniversary edition with a new preface, Berkeley: University of California Press, 2009.
Brouwer, L. E. J. ([1948] 1964). Consciousness, philosophy, and mathematics. In E. W. Beth, H. J. Pos, & J. H. A. Hollak (Eds.), *Proceedings of the Tenth International Congress of Philosophy* (pp. 1245–1249). Amsterdam: North-Holland. Reprinted in P. Benacerraf and H. Putnam (Eds.), *Philosophy of mathematics*, 1st ed., (pp. 78–84). Oxford: Blackwell.
Brouwer, L. E. J. ([1952] 1975a). Historical background, principles and methods of intuitionism. *South African Journal of Science, 49*, 139–146. Reprinted in Brouwer, L. E. J., *Collected works*, ed. by A. Heyting, (pp. 508–515). Amsterdam: North Holland, vol. 1.
Brouwer, L. E. J. (1975b). *Collected works* (vol. 1, A. Heyting, ed.). Amsterdam: North Holland.
Brouwer, L. E. J. (1981). *Cambridge lectures on intuitionism*, edited by D. van Dalen. Cambridge: Cambridge University Press.
Chomsky, N. (1972). *Studies in semantics in generative grammar*. The Hague: Mouton.
Chomsky, N. (1980). *Rules and representations. N.* York: Columbia University.
Chomsky, N. (1993). *Language and thought*. Wakefield, RI: Moyer Bell.
Chomsky, N. (1995). Language and nature. *Mind, 104*(413), 1–61.
Collins, H. M. (1983). An empirical relativist programme in the sociology of scientific knowledge. In K. D. Knorr-Cetina & M. Mulkay (Eds.), *Science observed: Perspectives in the social study of science* (pp. 85–100). London: Sage.
Collins, H. M., & Pinch, T. (1993). *The Golem: What everyone should know about science*. Cambridge: Cambridge University Press.
Conant, J., & Haugeland, J. (2000). Editor's introduction. In T. S. Kuhn (Ed.), *The road since structure: Philosophical essays, 1970–1993, with an autobiographical interview* (pp. 1–9). Chicago, IL: The University of Chicago Press.
De Groot, J. (1987). On the surprising in science and logic. *Review of Metaphysics, 40*(4), 631–655.

de Regt, H. W. (2017). *Understanding scientific understanding*. Oxford: Oxford University Press.

Diederich, W., Ibarra, A., & Mormann, T. (1989). Bibliography of the development of structuralism. *Erkenntnis, 30*(3), 387–407.

Dummett, M. (1977). *Elements of intuitionism*. Oxford: Clarendon Press.

Dummett, M. ([1963] 1978a). Realism (I). Lecture before the Oxford University Philosophical Society on 8 March 1963; reprinted in M. Dummett, *Truth and other enigmas* (pp. 145–165). London: Duckworth.

Dummett, M. (1978b). Preface. In M. Dummett (Ed.), *Truth and other enigmas* (pp. IX–LVIII). London: Duckworth.

Dummett, M. (1978c). Frege's distinction between sense and reference. In M. Dummett (Ed.), *Truth and other enigmas* (pp. 116–144). London: Duckworth.

Dummett, M. (1978d). Conocimiento práctico y conocimiento del lenguaje. *Anuario filosófico, 11*(1), 39–58.

Dummett, M. (1981a). *Frege: Philosophy of language* (2nd ed.). London: Duckworth (1st ed. 1973).

Dummett, M. (1981b). Frege and Wittgenstein. In I. Block (Ed.), *Perspectives on the philosophy of Wittgenstein* (pp. 31–42). Oxford: B. Blackwell.

Dummett, M. (1982). Realism (II). *Synthese, 52*(1), 55–112.

Dummett, M. (1991). *Frege: Philosophy of mathematics*. London: Duckworth.

Echeverria, J. (1990). Unidad de la Ciencia y concepción estructural. In W. J. Gonzalez (Ed.), *Aspectos metodológicos de la investigación científica. Un enfoque multidisciplinar* (2nd ed., pp. 173–188). Madrid-Murcia: Ediciones de la Universidad Autónoma de Madrid and Publicaciones de la Universidad de Murcia.

Frege, G. ([1892] 1967a). Über Sinn und Bedeutung. *Zeitschrift für Philosophie und philosophische Kritik, 100*, 25–50. Reprinted in Frege, G., *Kleine Schriften* (I. Angelelli, ed., pp. 143–162). Darmstadt: Wissenschaftliche Buchgesellschaft, 1967.

Frege, G. ([1918] 1967b). Der Gedanke. *Breiträge zur Philosophie des deutschen Idealismus, 1*, 58–77. Reprinted in Frege, G., *Kleine Schriften* (I. Angelelli, ed., pp. 342–362). Darmstadt: Wissenschaftliche Buchgesellschaft, 1967.

Frigg, R. (2006). Scientific representation and semantic view of theories. *Theoria: An International Journal for Theory, History and Foundations of Science, 55*(21), 49–65.

Geach, P. T. ([1958–59] 1970). Russell on meaning and denoting. *Analysis, 19*(3), 69–72. Reprinted in E. D. Klemke (Ed.), *Essays on B. Russell* (pp. 209–212). Urbana, IL: University of Illinois Press, 1970.

Geach, P. T. (1987). Reference and Buridan's law. *Philosophy, 62*, 7–15.
Geach, P. T., & Black, M. (Eds.) ([1952] 1960). *Translations from the philosophical writings of Gottlob Frege* (2nd ed.). Oxford: B. Blackwell.
Gonzalez, W. J. (1986a). *La Teoría de la Referencia. Strawson y la Filosofía Analítica*. Salamanca-Murcia: Ediciones Universidad de Salamanca and Publicaciones de la Universidad de Murcia.
Gonzalez, W. J. (1986b). El problema de la referencia en la Filosofía Analítica. *Estudio bibliográfico. Thémata, 3*, 169–213.
Gonzalez, W. J. (1987). The notion of "truth" in Wittgenstein's philosophy of mathematics. In P. Weingartner & G. Schurz (Eds.), *Logic, philosophy of science and epistemology* (pp. 419–423). Vienna: Hölder-Pichler-Tempsky.
Gonzalez, W. J. (1990a). Semántica anti-realista: Intuicionismo matemático y concepto de verdad. *Theoria: An International Journal for Theory, History and Foundations of Science, 5*(12–13), 149–170.
Gonzalez, W. J. (1990b). Progreso científico, autonomía de la Ciencia y realismo. *Arbor, 135*(532), 91–109.
Gonzalez, W. J. (1991a). Intuitionistic mathematics and Wittgenstein. *History and Philosophy of Logic, 12*(2), 167–183.
Gonzalez, W. J. (1991b). Mathematics as activity. *Daimon, 3*, 113–130.
Gonzalez, W. J. (1993). El realismo y sus variedades: El debate actual sobre las bases filosóficas de la Ciencia. In A. Carreras (Ed.), *Conocimiento, Ciencia y Realidad* (pp. 11–58). Zaragoza: SIUZ-Ediciones Mira.
Gonzalez, W. J. (1998). Prediction and prescription in economics: A philosophical and methodological approach. *Theoria: An International Journal for Theory, History and Foundations of Science, 13*(32), 321–345.
Gonzalez, W. J. (2004). Las revoluciones científicas y la evolución de Thomas S. Kuhn. In W. J. Gonzalez (Ed.), *Análisis de Thomas Kuhn: Las revoluciones científicas* (pp. 15–103). Madrid: Trotta.
Gonzalez, W. J. (2005). The philosophical approach to science, technology and society. In W. J. Gonzalez (Ed.), *Science, technology and society: A philosophical perspective* (pp. 3–49). A Coruña: Netbiblo.
Gonzalez, W. J. (2006). Novelty and continuity in philosophy and methodology of science. In W. J. Gonzalez & J. Alcolea (Eds.), *Contemporary perspectives in philosophy and methodology of science* (pp. 1–28). A Coruña: Netbiblo.
Gonzalez, W. J. (2008a). Evolutionism from a contemporary viewpoint: The philosophical-methodological approach. In W. J. Gonzalez (Ed.), *Evolutionism: Present approaches* (pp. 3–59). A Coruña: Netbiblo.

Gonzalez, W. J. (2008b). Rationality and prediction in the sciences of the artificial: Economics as a design science. In M. C. Galavotti, R. Scazzieri, & P. Suppes (Eds.), *Reasoning, rationality, and probability* (pp. 165–186). Stanford: CSLI Publications.

Gonzalez, W. J. (2010). *La predicción científica: Concepciones filosófico-metodológicas desde H. Reichenbach a N. Rescher*. Barcelona: Montesinos.

Gonzalez, W. J. (2011). Conceptual changes and scientific diversity: The role of historicity. In W. J. Gonzalez (Ed.), *Conceptual revolutions: From cognitive science to medicine* (pp. 39–62). A Coruña: Netbiblo.

Gonzalez, W. J. (2013a). The roles of scientific creativity and technological innovation in the context of complexity of science. In W. J. Gonzalez (Ed.), *Creativity, innovation, and complexity in Science* (pp. 11–40). A Coruña: Netbiblo.

Gonzalez, W. J. (2013b). Scientific prediction in the beginning of the 'historical turn:' Stephen Toulmin and Thomas Kuhn. *Open Journal of Philosophy, 3*(2), 351–357.

Gonzalez, W. J. (2013c). The sciences of design as sciences of complexity: The dynamic trait. In H. Andersen, D. Dieks, W. J. Gonzalez, T. Uebel, & G. Wheeler (Eds.), *New challenges to philosophy of science* (pp. 299–311). Dordrecht: Springer.

Gonzalez, W. J. (2014). On representation and models in Bas van Fraassen's approach. In W. J. Gonzalez (Ed.), *Bas van Fraassen's approach to representation and models in science* (pp. 3–37). Synthese Library, Dordrecht: Springer.

Gonzalez, W. J. (2015). *Philosophico-methodological analysis of prediction and its role in economics*. Dordrecht: Springer.

Gonzalez, W. J. (2020a). Novelty in scientific realism: New approaches to an ongoing debate. In W. J. Gonzalez (Ed.), *New approaches to scientific realism* (pp. 1–23). Boston/Berlin: De Gruyter. https://doi.org/10.1515/9783110664737-001.

Gonzalez, W. J. (2020b). Pragmatic realism and scientific prediction: The role of complexity. In W. J. Gonzalez (Ed.), *New approaches to scientific realism* (pp. 251–287). Boston/Berlin: De Gruyter. https://doi.org/10.1515/9783110664737-012.

Gonzalez, W. J. (2020c). Pragmatism and pluralism as methodological alternatives to monism, reductionism and universalism. In W. J. Gonzalez (Ed.), *Methodological prospects for scientific research: From pragmatism to pluralism* (pp. 1–18). Synthese Library, Cham: Springer.

Gonzalez, W. J. (2021). The relevance of language for scientific research. In W. J. Gonzalez (Ed.), *Language and scientific research.* (pp. 1–38). Cham: Palgrave Macmillan.
Gonzalez, W. J., & Arrojo, M. J. (2019). Complexity in the sciences of the Internet and its relation to communication sciences. *Empedocles: European Journal for the Philosophy of Communication, 10*(1), 15–33. https://doi.org/10.1386/ejpc.10.1.15_1.
Hacking, I. (1983). *Representing and intervening.* Cambridge: Cambridge University Press (reprinted 2005).
Hacking, I. (1999). *The social construction of what?* Harvard: Harvard University Press.
Howson, C. (2006). Scientific reasoning and the Bayesian interpretation of probability. In W. J. Gonzalez & J. Alcolea (Eds.), *Contemporary perspectives in philosophy and methodology of science* (pp. 31–45). A Coruña: Netbiblo.
Howson, C., & Urbach, P. (1989). *Scientific reasoning: The Bayesian approach.* La Salle, IL: Open Court. (This book has a second edition — Open Court, La Salle, IL, 1993 — and a third one in 2006.)
Irzik, G. (2001). Thomas Kuhn: The road since *Structure. Philosophy of Science, 68*(4), 573–575.
Janik, A., & Toulmin, S. E. (1973). *Wittgenstein's Vienna.* New York: Simon and Schuster (reprinted in 1996).
Jeffrey, R. C. (1975). Probability and falsification: Critique of the Popper program. *Synthese, 30*(1–2), 95–117.
Kellert, S. H., Longino, H. E., & Waters, C. K. (Eds.). (2006). *Scientific pluralism* (XIX Minnesota Studies in the Philosophy of Science). Minneapolis, MN: Minnesota University Press.
Kirkham, R. L. (1992). *Theories of truth.* Cambridge, MA: The MIT Press.
Kitcher, P. (2011). Scientific realism: The truth in pragmatism. In W. J. Gonzalez (Ed.), *Scientific realism and democratic society: The philosophy of Philip Kitcher.* Poznan Studies in the Philosophy of the Sciences and the Humanities (pp. 171–189). Amsterdam: Rodopi.
Kroon, F. W. (1985). Theoretical terms and the causal view of reference. *Australasian Journal of Philosophy, 63*(2), 143–166.
Kuhn, Th. S. (1962). *The structure of scientific revolutions.* Chicago: The University of Chicago Press, IL (2nd ed., 1970).
Kuhn, Th. S. ([1974] 1977). Second thoughts on paradigms. In F. Suppe (Ed.), *The structure of scientific theories* (pp. 459–482). Urbana, IL: University of

Illinois Press, 1974 (2nd ed., 1977). Reprinted in Kuhn, Th. S., *The essential tension* (pp. 293–319). Chicago: The University of Chicago Press.
Kuhn, T. S. (1990). Dubbing and redubbing: The vulnerability of rigid designation. In C. W. Savage (Ed.), *Scientific theories* (Minnesota Studies in the Philosophy of Science, Vol. XIV) (pp. 298–318). Minneapolis: Minnesota University Press.
Kuhn, Th. S. ([1983] 2000a). Commensurability, comparability, communicability. In P. D. Asquith & Th. Nickles (Eds.), *PSA 1982. Proceedings of the 1982 Biennial Meeting of the Philosophy of Science Association* (vol. 2, pp. 669–688). East Lansing, MI: Philosophy of Science Association. Reprinted in Th. S. Kuhn, *The road since* Structure (pp. 33–53). Chicago, IL: The University of Chicago Press.
Kuhn, Th. S. ([1987] 2000b), What are scientific revolutions? In L. Krüger, L. J. Daston, & M. Heidelberger (Eds.). *The probabilistic revolution. Vol. 1: Ideas in history* (pp. 7–22). Cambridge, MA: The MIT Press. Reprinted in Th. S. Kuhn, *The road since* Structure (pp. 13–32). Chicago, IL: The University of Chicago Press.
Kuhn, Th. S. ([1991a] 2000c). The road since *Structure.* in A. Fine, M. Forbes, & L. Wessels (Eds.), *PSA 1990. Proceedings of the 1990 Biennial Meeting of the Philosophy of Science Association* (vol. 2, pp. 3–13). Michigan, MI: Philosophy of Science Association, East Lansing (Presidential address delivered in 20 October 1990 in Minneapolis). Reprinted in Th. S. Kuhn, *The road since* Structure (pp. 90–104). Chicago, IL: The University of Chicago Press.
Kuhn, Th. S. ([1991b] 2000d). The trouble with the historical philosophy of science. Lecture delivered at Harvard University on 19 November 1991. Reprinted in Th. S. Kuhn, *The road since Structure. Philosophical essays, 1970–1993*, with an autobiographical interview, edited by James Conant and John Haugeland (pp. 105–120). Chicago, IL: The University of Chicago Press.
Ladyman, J. (2007). Scientific structuralism: On the identity and diversity of objects in a structure. *Aristotelian Society Supplementary Volume, 81*(1), 23–43.
Ladyman, J. (2011). Structural realism versus standard scientific realism: The case of phlogiston and dephlogisticated air. *Synthese, 180*(2), 87–101.
Lakatos, I. ([1970] 1978). Falsification and the methodology of scientific research programmes. In I. Lakatos & A. Musgrave (Eds.), *Criticism and the growth of knowledge* (pp. 91–196). Cambridge: Cambridge University Press. Reprinted in I. Lakatos, *The methodology of scientific research programmes.*

Philosophical papers, vol. 1, (pp. 8–101). Cambridge: Cambridge University Press.
Latour, B. (1988). *The pasteurisation of France*. Cambridge, MA: Harvard University Press.
Latour, B. (1993). *We have never been modern* (C. Porter, Trans.). Brighton: Harvester.
Latour, B., & Woolgar, S. (1986). *Laboratory life: The social construction of scientific facts* (2nd ed.). Princeton, NJ: Princeton University Press.
Laudan, L. (1984). Realism without the real. *Philosophy of Science, 51*(1), 156–162.
McGinn, C. (1982). Realist semantics and content-ascription. *Synthese, 52*(1), 115–127.
Moulines, C. U. (1982). *Exploraciones metacientíficas*. Madrid: Alianza Editorial.
Moulines, C. U. (1990). Frege, la verdad y el desarrollo científico. In W. J. Gonzalez (Ed.), *Aspectos metodológicos de la investigación científica. Un enfoque multidisciplinar* (2nd ed., pp. 97–114). Madrid-Murcia: Ediciones de la Universidad Autónoma de Madrid and Publicaciones de la Universidad de Murcia.
Nagel, E. (1961). *The structure of science. Problems in the logic of scientific explanation*. New York: Harcourt, Brace and World.
Niiniluoto, I. (1984). *Is science progressive?* Dordrecht: Reidel.
Niiniluoto, I. (1993). The aim and structure of applied research. *Erkenntnis, 38*(1), 1–21.
Niiniluoto, I. (1995). Approximation in applied science. *Poznan Studies in the Philosophy of the Sciences and the Humanities, 42*, 127–139.
Niiniluoto, I. (1999). *Critical scientific realism*. Oxford: Clarendon Press.
Peters, D. (2014). What elements of successful scientific theories are the correct targets for 'selective' scientific realism? *Philosophy of Science, 81*(3), 377–397.
Psillos, S. (1994). P. Horwich: World changes. Thomas Kuhn and the nature of science — P. Hoyningen-Huene: Reconstructing scientific revolutions. Thomas Kuhn's philosophy of science. *British Journal for the Philosophy of Science, 45*(3), 923–926.
Psillos, S. (1999). *Scientific realism. How science tracks truth*. London: Routledge.
Putnam, H. (1973). Meaning and reference. *Journal of Philosophy, 70*(19), 699–711. Reprinted in S. P. Schwartz (Ed.), *Naming, necessity, and natural kinds* (pp. 118–132). Ithaca: Cornell University Press, 1977.
Putnam, H. (1975). Language and reality. In H. Putnam, *Mind, language and reality* (pp. 272–290). Cambridge, MA: Cambridge University Press.

Putnam, H. (1975–76). What is "realism." *Proceedings of the Aristotelian Society, 76*, 177–194.

Putnam, H. (1978a). Realism and reason. In H. Putnam, *Meaning and the moral sciences* (pp. 123–140). London: Routledge and K. Paul.

Putnam, H. (1978b). Reference and Understanding. In H. Putnam, *Meaning and the moral sciences* (pp. 97–119). London: Routledge and K. Paul.

Putnam, H. (1983a). *Realism and reason*. Cambridge: Cambridge University Press.

Putnam, H. (1983b). Reference and Truth. In H. Putnam, *Realism and reason* (pp. 69–86). Cambridge: Cambridge University Press.

Putnam, H. (1990). A defense of internal realism. In H. Putnam, *Realism with a human face*, edited by James Conant (pp. 30–42). Harvard University Press, Cambridge.

Putnam, H. (1992). A theory of reference. In H. Putnam, *Renewing philosophy* (pp. 35–59). Cambridge, MA: Harvard University Press.

Reichenbach, H. (1938). *Experience and prediction. An analysis of the foundations and the structure of knowledge*. Chicago, IL: The University of Chicago Press.

Rescher, N. (1977). *Methodological pragmatism: A systems-theoretical approach to the theory of knowledge*. Oxford: Blackwell; New York: New York University Press.

Rescher, N. (1992). *A system of pragmatic idealism*. Vol. I: *Human knowledge in idealistic perspective*. Princeton: Princeton University Press.

Rescher, N. (1998a). *Predicting the future*. New York: State University of New York Press.

Rescher, N. (1998b). *Communicative pragmatism and other philosophical essays on language*. Lanham, MD: Rowman and Littlefield.

Rescher, N. (2012). *Pragmatism: The restoration of its scientific roots*. New Brunswick: Transaction Publishers.

Rescher, N. (2014). *The pragmatic vision: Themes in philosophical pragmatism*. Lanham, MD: Rowman and Littlefield.

Rescher, N. (2020). Methodological pragmatism. In W. J. Gonzalez (Ed.), *Methodological prospects to scientific research: From pragmatism to pluralism* (pp. 69–80). Synthese Library, Dordrecht: Springer.

Resnik, M. (1987). You can't trust an ideal theory to tell the truth. *Philosophical Studies, 52*, 151–160.

Rolin, K. (2015). Values in science: The case of scientific cooperation. *Philosophy of Science, 82*(2), 157–177.

Schmitt, F. F. (Ed.). (2004). *Theories of truth*. Oxford: Blackwell.

Simon, H. A. (1996). *The sciences of the artificial* (3rd ed.). Cambridge, MA: The MIT Press.

Sneed, J. (1971). *The logical structure of mathematical physics*. Dordrecht: Reidel.

Strawson, P. F. ([1950] 1971a). On referring. *Mind, 59*(235), 320–344. Reprinted in P. F. Strawson, *Logico-linguistic papers* (pp. 1–27). London: Methuen.

Strawson, P. F. ([1970] 1971b). *Meaning and truth* (An inaugural lecture at the University of Oxford, delivered on 5 November 1969). Oxford: Clarendon Press, 1970. Reprinted as Strawson, P. F., Meaning and truth. In P. F. Strawson., *Logico-linguistic papers* (pp. 170–189). London: Methuen.

Strawson, P. F. (1986). Reference and its roots. In L. E. Hahn & P. A. Schilpp (Eds.), *The philosophy of W. V. Quine* (The Library of Living Philosophers) (pp. 519–532). La Salle, IL: Open Court.

Suppe, F. ([1974] 1977). The search for philosophic understanding of scientific theories. In F. Suppe (Ed.), *The structure of scientific theories* (pp. 1–241). Urbana, IL: University of Illinois Press.

Suppes, P. (1962). Models of data. In E. Nagel, P. Suppes, & A. Tarski (Eds.), *Logic, methodology and philosophy of science* (pp. 252–261). Stanford: Stanford University Press.

Suppes, P. (1967). What is a scientific theory. In O. Morgenbesser (Ed.), *Philosophy of science today* (pp. 55–67). New York: Basic Books.

Suppes, P. (1984 [1985]). *Probabilistic metaphysics*. Oxford: B. Blackwell (reprinted in 1985).

Tiropanis, T., Hall, W., Crowcroft, J., Contractor, N., & Tassiulas, L. (2015). Network science, Web science, and Internet science. *Communications of ACM, 58*(8), 76–82.

Toulmin, S. (1961). *Foresight and understanding. An enquiry into the aims of science*. Bloomington: Indiana University Press, IN.

Toulmin, S. E. (1953 [1957]). *The philosophy of science. An introduction*. London: Hutchinson University Library (3rd reprint, 1957).

Toulmin, S. E. (1970). Does the distinction between normal and revolutionary science hold water? In I. Lakatos & A. Musgrave (Eds.), *Criticism and the growth of knowledge* (pp. 39–47). Cambridge: Cambridge University Press.

Tuomela, R. (1979). Putnam's realism. *Theoria: An International Journal for Theory, History and Foundations of Science, 45*, 114–126.

Tuomela, R. (1985). *Action, science and reality*. Dordrecht: Reidel.

Tuomela, R. (1987). Scientific realism and best explanation. In P. Weingartner & G. Schurz (Eds.), *Logic, philosophy of science and epistemology* (pp. 196–202). Vienna: Hölder-Pichler-Tempsky.

Tuomela, R. (1990). Causal internal realism. In G. Pasternak (Ed.), *Philosophie und Wissenschaften* (pp. 165–179). Frankfurt: P. Lang.

van Fraassen, B. C. (2008). *Scientific representation: Paradoxes of perspective*. New York: Oxford University Press.

Wittgenstein, L. (1922). *Tractatus logico-philosophicus* (bilingual edition) (German-English). London: Kegan Paul.

Wittgenstein, L. (1953 [1976, 2001]). *Philosophische Untersuchungen* (G. E. M. Anscombe and R. Rhees, ed.). Oxford: B. Blackwell. Translated into English by G. E. M. Anscombe: *Philosophical investigations*. Oxford: B. Blackwell, 1953 (reprint of the English text with indexes, 1976); 3r bilingual edition (German-English), 2001.

Woolgar, S. (1981). Critique and criticism: Two readings of ethnomethodology. *Social Studies of Science, 11*(4), 504–514.

Woolgar, S. (1988a). *Science: The very idea*. London: Tavistock.

Woolgar, S. (Ed.). (1988b). *Knowledge and reflexivity: New frontiers in the sociology of knowledge*. London: Sage.

Wright, C. (1981). Critical study: Dummett and revisionism. *Philosophical Quarterly, 31*(122), 47–67.

Wright, C. (1987). *Realism, meaning and truth*. Oxford: B. Blackwell.

3

On the Role of Language in Scientific Research: Language as *Analytic, Expressive, and Explanatory* Tool

Ladislav Kvasz

The analysis of the relationship between language and scientific research can be approached from different perspectives. One such perspective is offered by history of science. It consists in exploring the development of our ability to measure certain magnitudes, construct models of particular phenomena, present laws explaining these phenomena, and theoretically justify these laws. A fact that probably cannot be denied is that today we can measure, model, explain, and justify much more than our ancestors in the seventeenth or eighteenth century. This does not mean that we are wiser; our superiority is related to the fact that we have much better instruments. One of these instruments that substantially influenced scientific thought is *language*. Language influences scientific thought in many ways — by means of appropriate metaphors language enhances its depth, by means of fine distinctions language increases its exactness, and often it opens to thought completely new possibilities. To be able to describe the ways in which language influences thinking, I introduce the

L. Kvasz (✉)
Institute of Philosophy, The Czech Academy of Sciences,
Prague, Czech Republic

W. J. Gonzalez (ed.), *Language and Scientific Research*,
https://doi.org/10.1007/978-3-030-60537-7_3

concept of *potentialities of language*. This notion refers to the new possibilities that language opens for thought. The aim of the paper is to describe the potentialities of the language of physics.

1 An Overview of the Potentialities of the Language of Mathematics

The potentialities of language were introduced in the paper *Changes of Language in the Development of Mathematics* (Kvasz 2000, 50) as a tool for analyzing the development of mathematics. I originally distinguished four potentialities, and later in the paper *Language in Change* (Kvasz 2012) I added two other, which led to the following list:

1. *Logical power*, which shows how complex formulas the language allows us to prove, formulas which were in the previous stages of development unprovable.
2. *Expressive power*, which shows what new things the language allows us to express, things which in the previous stages of development defied expression.
3. *Methodical power*, which shows what new methods the language enables us to introduce there, where on the previous stages of development we saw only several unrelated tricks.
4. *Integrative power*, which shows what sort of unity and order the language enables us to see there, where in the previous stages we perceived just unrelated particular cases.
5. *Explanatory power*, which shows how the language allows us to explain the failures which occurred in the previous stages, failures that were previously incomprehensible.
6. *Constitutive power*, which shows how the language, by transgressing the rules of its own syntax, enables us to constitute some radically new kinds of objects.

The development of the language of mathematics consists in the increase of the logical and expressive power: the language allows us to prove more and more theorems and to describe a still wider area of phenomena.

Gradually increases its methodical and integrative power: the language allows us to create more effective methods and provides a more comprehensive view of the world. Similarly increases also the explanatory and the constitutive power: the language allows us to understand our past failures and to introduce new kinds of objects. The process of gradual increase of the potentialities of language of mathematics is described in (Kvasz 2012). My aim here is to transfer the notion of potentialities of language from mathematics to physics.

But before we turn to this problem, we need to distinguish the main stages in the development of the language of physics. This is important, because only on the background of an unprejudiced, balanced, and sufficiently complex picture of the development of physics it is possible to discern the pattern of development of the potentialities of the language. The increase of a particular potentiality, such as the expressive or the explanatory power of language, can be most easily recognized by confronting the linguistic frameworks of successive developmental stages. Therefore, it is important not to exclude some of the stages and not to mix stages belonging to different lines of development.

2 History of Physics from a Linguistic Point of View

The first step of the linguistic reconstruction of the history of physics is to identify its stages. In principle we accept the main stages introduced in the literature on history of science: *Newtonian physics*, *classical physics*, and *quantum physics*. Unlike the standard historical accounts, however, we place emphasis on language and characterize the particular developmental stages of physics by means of their linguistic framework. Therefore, we will deviate slightly from the standard interpretation.

As a first difference we suggest to separate the physics of the second half of the eighteenth century from the general framework of Newtonian physics and single it out as a separate stage under the name *physics of continua and fluids*. Thus what is usually termed Newtonian physics, we propose to divide into two stages, and call them *Newtonian physics* (in a narrow sense) and *physics of continua and fluids*. The physics of continua and fluids is a set of theories such as the theory of the vibrating strings, the theory of elasticity, hydrodynamics, the theory of heat conduction, phlogiston theory of combustion, or theory of the electric and magnetic fluid. What unites these theories and brings them into a certain conflict with Newtonian physics is that they are abandoning the corpuscular ontology and interpret the observed phenomena by means of the motion of a material continuum or an imponderable fluid. Instead of Newton's second law, which is an ordinary differential equations describing the motion of a particle, these theories *hypothetically single out an infinitesimal element* of the continuum or fluid, write down the *balance equation of the hypothetical forces* by which the neighboring elements act on this element and derive a *partial differential equation*, that describes the studied process. From a linguistic point of view these theories differ quite significantly from Newton's approach and at the same time manifest a sufficiently high degree of unity to be considered a separate stage in the development of physics.

In a similar way, we suggest to distinguish as a separate stage also physics of the first half of the nineteenth century, which we propose to call *physics of atoms and energies*. It is a period of crisis of physics of continua and fluids when the different continua disintegrated into atoms and the fluids are being gradually replaced by the concept of energy. Physics of atoms and energies is usually included into classical (i.e. pre-quantum) physics. Thus what is usually referred to as the period of classical physics, we propose to divide into two stages, which we suggest to call *physics of atoms and energies* and *physics of fields*. We propose to discriminate physics of atoms and energies as a separate stage, because at the linguistic level it clearly differs from physics that preceded it as well as from that which came after it. In the history of physics, we thus distinguish the following five linguistic frameworks.

2.1 Newtonian Physics

In his *Philosophiae Naturalis Principia Mathematica* (Newton 1687) Isaac Newton described interaction among bodies by means of action of forces.

In addition to *contact forces* that act in collisions, and *forces of friction*, he introduced *forces acting at a distance*. In action at a distance it is not necessary for the acting bodies to touch each other. Forces acting at a distance are the gravitational forces between planets, described by Newton's law of universal gravitation, or the electrostatic forces between charged bodies described by Coulomb's law that was discovered by Charles Auguste de Coulomb in 1784.

2.2 Physics of Continua and Fluids

In Newtonian physics the equations motion are ordinary differential equations. In his paper *De motu nervi tensi* (Taylor 1713) dedicated to the description of the vibrating string Brook Taylor has for the first time singled out an element of the string and examined the forces acting on it from side of the surrounding elements. This allowed him to determine the simplest form of the vibrations. Leonard Euler turned the trick of Taylor into a general method when he formulated in his work *A discovery of a new principle of mechanics* (Euler 1750) the principle that the differential equations describing the motion of a free body remain valid also when we use them to describe the motion of an element of the body or an element of a fluid. Thus physics of continua and fluids was born, according to which a body is composed of elements which are *singled out hypothetically*. These elements have the same properties as the continuum as such (i.e. the same density, elasticity, hardness) but they are small enough to allow for a limit to differentials. From a mathematical point of view thus alongside ordinary differential equations, that describe the motion of the body as a whole, partial differential equations appear as the equation of the vibrations of a homogenous string (Euler 1748), the equation of the flow of fluids (Euler 1755), and the equation of the conduction of heat (Fourier 1822). These equations describe *the spread of action in a continuous medium*. Physics of continua and fluids was able in this way to describe a broad class of phenomena and led to the emergence of the mechanistic world view.

2.3 Physics of Atoms and Energies

Physics of atoms and energies was born from the crisis of the theories of continua and fluids. This crisis was caused by progress in several disciplines. In chemistry it was the discovery that air is not a simple elastic continuum, as it was described by physics of continua and fluids, but is a mixture of different substances. In 1755 Joseph Black discovered carbon dioxide, in 1766 and Henry Cavendish discovered hydrogen. These discoveries led to the publication of the *Traité Elémentaire de Chimie* (Lavoisier 1789) where Antoine Laurent Lavoisier presented his oxidation theory of combustion, which replaced the phlogiston theory and gave birth to the concept of a chemical element. Progress in calorimetry led to Joule's experiments to determine the mechanical equivalent of heat, which in 1843 showed that heat is not a fluid. Lavoisier discarded the phlogiston in 1789, James Prescott Joule the caloric in 1843 and Albert Einstein brought an end to the last fluid — the luminiferous ether in 1905. A parallel development of the theory of materials (such as optical glass, or industrial steel) during the first half of the nineteenth century brought physics beyond the possibilities of the description of matter as a continuous medium and led to a fundamental change in our understanding of the structure of matter.

The result of all these changes was that in the mid nineteenth century the hypothetical postulation of elements of the mathematical continuum was replaced by the study of the physical composition of materials and of the processes that take place in them by means of empirical methods. The increase of the accuracy of experimental methods allowed physics to make a step beyond the macroscopic level of description of an element of the continuum and to start to explore the real microscopic structure of matter. Similarly, the various fluids, that were postulated as embodiments of macroscopic processes (such as heat flow, combustion, electric current), were disintegrated and turned into collective movements of real particles (such as atoms, molecules, ions or electrons).

2.4 Field Theories

Field theory grew out of the work Michael Faraday who introduced lines of force in order to visualize the action of electric charges and currents in his *Experimental Researches in Electricity* (Faraday 1831). Faraday used these lines for instance in the description of electromagnetic induction that he discovered. Most physicists did not take the lines of force seriously, seeing in them merely a heuristic device. That the lines of force are not a mere tool, enabling us to visualize processes that take place in experiments, but that they express an important physical content was first understood by Maxwell. In his paper *On the Phisical Lines of Force* (Maxwell 1861) he rewrote Faraday's qualitative descriptions into mathematical form, and passed from lines of force to the notion of a field. Maxwell attributed to lines of force an objective reality. His theory of the electric and magnetic fields formulated in *A Treatise on Electricity and Magnetism* (Maxwell 1873) was further developed by Hendrik Antoon Lorentz, who introduced a description of the action of the field on matter in motion. Nevertheless, Lorentz still strived to reconcile field theory with principles of physics of atoms and energies. Therefore, he introduced laws describing the contraction of atoms in motion, now called the Lorentz's transformations. Gradually it became clear that the reconciliation with physics of atoms and energies is impossible, and in 1905 Einstein concluded in his paper *Zur Elektrodynamik bewegter Körper* (Einstein 1905a) that field theory requires a new interpretation of the fundamental notions of space and time, that goes far beyond the possibilities of physics of atoms and energies. Thus while Maxwell still tried to explain the generation of the field by models of motion of the particles of the eather, Einstein postulated the existence of the electromagnetic field independently of any such substance.

2.5 Quantum Physics

The first work in which the hypothesis of quanta appeared was the paper of Max Planck *Zur Theorie des Gesetzes der Energieverteilung im Normalspektrum* (Planck 1900). It was published in 1900 and devoted to the problem of black body radiation. Previous attempts to describe black body radiation have led to divergent formulas. Planck reached a satisfactory result at the cost of the supposition that the black body does not radiate continuously, as it is required by Maxwell's theory, but in small portions, which he called quanta. Einstein used Planck's hypothesis in his paper *Über einen die Erzeugung und Verwandlung des Lichtes betreffenden heuristischen Gesichtspunkt* (Einstein 1905b) to explain of the photoelectric effect and Niels Bohr has built the quantum hypothesis into the foundations of his theory of atoms in his paper *On the Constitution of Atoms and Molecules* (Bohr 1913). Bohr and Einstein assumed that the quanta of radiation are real and so in addition to atoms and electrons they postulated a further kind of objects: *quanta of radiation*. The next step in the development of quantum mechanics occurred in 1923 when Louis de Broglie reached in his PhD dissertation *Recherches sur la théorie des quanta* the conclusion that the quantum hypothesis concerns not only radiation but similarly as Planck attached to continuous waves of radiation discrete quanta, we can also attach to discrete particles of matter continuous waves. The quantum hypothesis has thus become a universal principle, applying to all phenomena. De Broglie's work was in rapid succession followed by works of Werner Heisenberg, Max Born, Ernst Pascual Jordan, Erwin Schrödinger, Paul Adrien Maurice Dirac, and Wolfgang Pauli, until in 1927 John von Neumann proposed the standard formulation of quantum mechanics, based on the concept of Hilbert spaces in his *Mathematische Begründung der Quantenmechanik* (von Neumann 1927).

3 Description of the Potentialities of the Language of Physics

After we have outlined in the second part of this paper the basic linguistic frameworks established in the development of physics, we can turn to the problem of identifying the particular potentialities of language. We will focus on the transitions between different frameworks and will try to find changes that can be interpreted as growth of the particular potentialities.

In doing so, of course, besides the illustrations that we have chosen it is possible to find several others that demonstrate the increase of the same potentiality, thus the following examples are by no means the only possible ones.

3.1 The Analytic Power of the Language of a Physical Theory

Transferring the notion of the logical power of language from mathematics to physics is not difficult. In physics it will be more appropriate to call it *analytic power of language*, and put it into relation with derivation of laws of nature rather than with proofs of theorems. So I propose to characterize the analytic power of the language of physics by the system of all formulas that can be analytically derived in a particular language. Thus with the term analytic power of language I will denote the *ability of language of a particular physical theory to derive a new type of relations* between quantities that in the previous stages could not be derived.

As a paradigmatic example illustrating the analytic power of the language of Newtonian physics we can take the *derivation of the law of universal gravitation from Kepler's laws*. For Kepler the elliptical shape of planetary orbits was an empirical fact. He discovered it by analyzing a large number of astronomical data. In the linguistic framework of Newtonian physics, it is possible from Kepler's laws derive the law of universal gravitation. From Kepler's second law it follows that the gravitational force is a centripetal force and from Kepler's third law follows its inverse proportionality to the square of distance. This derivation is the main result of the first book of Newton's *Principia*.

A similarly significant success of physics of continua and fluids was Fourier's *derivation of the equation of heat conduction*, published in the book *Théorie analytique de la chaleur* (Fourier 1822). Fourier derived his equation assuming that heat has the form of a fluid called caloric, and the equation describes its flow through the medium. Some twenty years after the publication of Fourier's book Joule showed that no such fluid exists. However, although we have abandoned caloric, Fourier's equation describes with high precision the thermal phenomena. Its derivation by

Fourier illustrates the analytic power of the language of physics of continua and fluids.

The analytic power of the language of physics of atoms and energies illustrates the *derivation of the equation of motion of the incompressible viscous fluid* given by Claude-Louis Navier in the paper *Mémoire sur les lois du mouvements des fluids* (Navier 1822). Although in the equation only quantities that characterize the fluid as a continuum occur, Navier derived his equation on the assumption that the fluid consists of molecules, and the forces of interaction between molecules are proportional to their relative velocity. The idea of molecules occurred in physics of continua and fluids as *a mere trick* that made possible the derivation of the equation of motion of the viscous fluid (just like later in the derivation of the law of the black body radiation Planck will introduce the quanta, or in the derivation of the equation of the vibrating string Taylor introduced the intra-material forces). The molecules are something fundamentally different from the elements, into which a continuum was divided. The properties of the continuum are *not* transferred onto the molecules; molecules do not have hardness or elasticity equal to the hardness and elasticity of the macroscopic body. On the contrary, from the features of the body the properties of the molecules are derived. The molecules constituting the liquid are not *hypothetical formal elements* but physically real particles of matter accessible, at least in principle, to empirical scrutiny. For example, the question of how big is the volume element dV of water does not make sense, while the question of how big is a water molecule makes sense, even though at times of Navier it was not possible to find the answer.

The analytic power of the language of field theory can be illustrated by the *discovery of Maxwell's displacement current*. After rewriting in a mathematical form all known facts about the electric and magnetic fields, Maxwell realized that the resulting equations are unbalanced. A varying magnetic field created an electric field (Faraday's law of electromagnetic induction), but a varying electric field had no analogous effect. Driven by the idea of symmetry Maxwell postulated the existence of a magnetic effect of a varying electric field. This effect was not yet observed because its registration requires very specific conditions, which were impossible to hit upon by mere trial and error. When he introduced into the equations an additional element, he found out that the equations have a solution in the form of electromagnetic waves. Maxwell published his discovery in 1873 and in 1886 Heinrich Hertz experimentally verified the existence of electromagnetic waves. Maxwell's discovery illustrates the analytic power of language, because it was the transcription of the experimental data into a mathematical *language* that revealed a gap in these data and after filling in this gap it became possible to *derive* the existence of electromagnetic waves.

3.2 Expressive Power of the Language of a Physical Theory

The expressive power of language can be defined as the ability of language to represent some aspect of nature that at previous stages of development defied linguistic expression. In the history of physics there are several phenomena, which for a long time were unknown or defied description. Only by means of the language of the new theory it became possible to describe them. To illustrate the expressive power of the language of Newtonian physics we can take *friction.* Cartesian physics described interaction between bodies by means of the law of conservation of the quantity of motion. This made it impossible to describe friction, because friction turns a certain amount of mechanical motion into heat and thus the Cartesian law of conservation of the quantity of motion is violated. Friction is a phenomenon which Cartesian physics cannot describe in a quantitative way. In Newtonian physics friction is simply another term on the right hand side of the equation of motion.

One of the few mistakes in Newton's *Principia* was the derivation of the speed of sound. The correct value of this speed was derived by Pierre Simon Laplace in his paper *Sur la vitesse du son dans l'air et dans l'eau* (Laplace 1816), when he realized that in a sound wave air undergoes not an isothermal compression, as Newton implicitly assumed. During compression in the sound wave the temperature of the air rises and this increase of temperature causes an increase of the speed of the wave. The theoretical justification of Laplace's derivation was given by Siméon Denis Poisson in his paper *Sur la vitesse du son* (Poisson 1823) in the framework of the theory of caloric by introducing the distinction between an isothermal and an adiabatic process. When we imagine an element of air as a sponge imbued with caloric, the *isothermal compression* of air corresponds to squeezing the sponge in such a way, that the excess of the caloric leaves the sponge. The compressed sponge has smaller volume, and in this smaller volume a smaller amount of caloric is contained. The excessive caloric must be transported away. When the sponge is squeezed too quickly, and in a sound wave we are dealing with alternating compression and expansion taking place 1000 times per second, the caloric does not manage to leave the sponge and is compressed with it. Squeezing the caloric, i.e. increasing the amount of caloric in a given volume, is nothing else than an increase of temperature. Poisson realized that in sound waves the compressions take place so quickly that the caloric is compressed together with the air and the isothermal character of the process is broken. We see how the idea of heat as a fluid allows us to represent the thermal processes accompanying the vibration of air in a sound wave. The ability to express the *difference between an isothermal and an adiabatic process* can be thus taken as an illustration of the expressive power of the language of physics of continua and fluids.

An example illustrating the expressive power of the language of physics of atoms and energies is the *determination of atomic mass of chemical elements*. Atoms of each chemical element have a particular mass that is characteristic for this element (and thus by dividing a chemical element into parts we cannot obtain arbitrarily small fragments, but there is a smallest fragment of each element). This fact cannot be expressed in the language of physics of continua and fluids. From the point of view of the language of physics of continua matter is a homogeneous continuum, and so the concept of atomic mass does not make sense.

To illustrate the expressive power of the language of field theory we can take the *description of Lorentz contraction*. Proponents of classical physics were not aware of the fact that the length of a body may vary with speed. The language they used did not allow expressing such a variation. Only in the linguistic framework of field theory, that postulated the invariance of the speed of light, was it possible to introduce the notion of local time and to define the length of a moving body using this notion. After this refinement of conceptual framework it turned out that the length of a body moving with high velocity changes as a function of its velocity. The description of a hitherto inexpressible phenomenon illustrates the increase in the expressive power of the language of physics.

3.3 Methodical Power of the Language of a Physical Theory

To transfer the methodical power of language from mathematics to physics is not as straightforward as in the previous two cases. In mathematics the methodical power of language is the ability of the language to replace random, *ad hoc* tricks by systematic methods. In physics I propose to relate the methodical power of language to *new methods of measurement.* These methods are often based on the ability of language to express a relation between quantities that were previously unrelated. Such an increase of the methodical power of language is often accompanied with the introduction of a new *fundamental constant* that is the means that enables us to correlate quantities that were at previous stages of the development unrelated. The different fundamental constants are a striking feature of the particular theory that makes it easier to identify the methodical power of its language.

As an example illustrating the methodical power of language of Newtonian physics is the "weighting of the Earth." In 1798 Henry Cavendish, using a fine torsion balance, succeeded in measuring the force of attraction between two big marble spheres (and thus to determine the value of the *gravitational constant* κ that appeared in Newton's law of universal gravitation). When this small force is compared with the weight of the spheres, i.e. with the force with which they are attracted to the Earth, it is possible to determine the mass of the Earth. Cavendish's measurement is therefore often called the "weighting of the Earth."

A fundamental constant that illustrates the methodical power of the language of physics of atoms and energies is *Boltzmann's constant* κ. This appears in a number of relations, that determine the distribution functions of different quantities in a gas, and thus allows us to move from a microscopic description at the molecular level to the quantities that are accessible to measurement at the macroscopic level. Therefore, the relation of this constant to the methodical power of the language of statistical physics is obvious.

The fundamental constant that constitutes the methodical power of the language of field theory is the *speed of light c*. In previous linguistic frames the speed of light depended on the relative velocities of the source and the detector of light, and therefore it was not constant. In field theory, on the other hand, the speed of light is constant (it is one of the postulates of relativity theory). This constant allows us to connect the energy E and the mass m of a particle using Einstein's formula $E = mc^2$. This formula lies at the basis of methods that use the measurement of one of these variables to determine the other. For example, the binding energy of nuclear forces can be determined from the mass defect of the atoms entering a nuclear reaction.

3.4 Integrative Power of the Language of a Physical Theory

The integrative power of the language can be transferred from mathematics to physics without any problem. An illustration of the integrative power of the language of Newtonian physics is the *unification of terrestrial and celestial mechanics.* We can see a predecessor of this unification in Galileo Galilei who observed in his telescope mountains on the surface of the moon and concluded from this that the moon is composed of the same substance as Earth (Galileo 1610). However, for Galileo the unification of terrestrial and celestial physics was synthetic, it was a result of observations. Galileo did not have at his disposal a linguistic framework that would allow expressing this unity formally. This was done by Newton; whose law of universal gravitation allows us to calculate trajectories of comets as well as the fall of an apple from the tree. Thus, the unification of celestial and terrestrial mechanics is achieved by Newton by means of language, and this unification is thus an illustration of the integrative power of the language of Newtonian physics.

The integrative power of the language of physics of continua and fluids found its manifestation in the *mechanistic world view.* The mechanistic world view is often attributed to Newton, but Newtonian physics does not have the necessary repertoire of theoretical tools that would integrate into a single picture of the world the description of mechanical, acoustic, thermodynamic, electrical, and optical phenomena. Therefore, the attribution of the mechanistic world view to Newton is a mistake. The mechanistic picture of the world is the manifestation of the integrative power of the language of physics of continua and fluids. It is a picture in which all phenomena from mechanical to optical ones find a uniform representation as a mechanical movements or a continuum or a fluid.

The integrative power of the language of physics of atoms and energies is expressed in the *law of conservation of energy*. Phenomena that the previous linguistic framework represented separately and even fixed this separation by postulating a different kind of fluid for each sort of phenomena are now integrated into a single framework. The unity of the mechanistic world view was a formal unity; it was based on the fact that the different physical phenomena have been described in the same way: by *postulating* a fluid or a continuum and a *describing* its motion by means of a differential equation. Each region, however, retained its specificity thanks to the special fluid. Thermodynamics described the motion of the caloric, electrodynamics the motion of the electric fluid, and optics the oscillations of the luniniferous ether. Physics of atoms and energies discarded these fluids (in the case of the caloric or the phlogiston), or atomized them (in the case of the electric fluid) and afterwards integrated all phenomena into a single coherent framework. Mechanical, acoustic, thermal, electrical, optical, and chemical phenomena — they all are processes of transformation of energy.

An illustration of the integrative power of the language of field theory is the *unification of electrodynamics and optics*, which is one of the great unifications in the history of science, when two areas of phenomena which were formerly unrelated turned out to be just different aspects of the same reality. After Maxwell discovered that electromagnetic waves are solutions of the equations of the electromagnetic field, he calculated the speed at which these waves propagate. As a result he received a number equal to the speed of light. Therefore, he hypothesized that light is an electromagnetic wave. Further research fully confirmed this hypothesis.

3.5 Explanatory Power of the Language of a Physical Theory

In mathematics I characterized the explanatory power of language by its ability to *explain the failures* of the linguistic framework of the previous developmental stage. Although using a language we can explain many other phenomena, the explanation of a failure of the previous framework convincingly illustrates the increase of explanatory power. A failure of a linguistic framework is a fact that is incomprehensible in terms of the framework itself, and thus an explanation of this failure is an unequivocal sign of the increase of the explanatory power. In physics, it is not difficult to find examples of explanatory power of language.

Newtonian physics could not explain *the regularity of the solar system* and Newton resorted in his explanation to the hypothesis of divine intervention. To use divine intervention in the explanation of physical phenomena can be seen as an implicit acknowledgment of the failure of Newtonian physics. For physics of continua and fluids is not difficult to explain this regularity. It is sufficient to assume that the solar system was formed by condensation from a dust cloud as Immanuel Kant assumed in his *Allgemeine Naturgeschichte und Theorie des Himmels* (Kant 1755), or was created by separation from the Sun as Laplace assumed in his *Exposition du systéme du monde* (Laplace 1796). It can be objected that the explanation of the regularity of the solar system by divine intervention or by a rotating dust cloud are equally good, because the dust cloud is as inaccessible to empirical investigation as a divine intervention. Even if that is true, we have to remember that every explanation that does not move in a circle must be based on assumptions that are no longer explained. The difference between explaining the regularity of the solar system by God's intervention and the explanations of Kant or Laplace is that the assumptions of Kant and Laplace are formulated in the language of physics. Newton's explanation, in contrast, is an explanation of a phenomenon expressed in the language of physics by a cause that is not expressible in that language. So there is an increase of explanatory power of language in passing from Newton to Kant and Laplace.

Irreversibility is a phenomenon that occurred already in the linguistic framework of physics of continua and fluids, namely in Fourier's theory of heat conduction. But as long as physicists believed in the existence of the caloric, the irreversibility of the thermal processes did not constitute a problem — it was simply one of the particularities of this new substance. Nevertheless, when Joule established that heat is the energy of motion of the microscopic particles of matter, the irreversibility of the thermal processes started to contradict the reversibility of the laws describing the motion on the microscopic level. This objection formulated by the Austrian physicist Johann Josef Loschmidt in 1876. A conceptual clarification of irreversibility was offered by Ludwig Eduard Boltzmann in 1877, and it illustrates the explanatory power of language of physics of atoms and energies.

3.6 Constitutive Power of the Language of a Physical Theory

I introduced the notion of the constitutive power of language of mathematics in order to be able to characterize changes such as the introduction of complex numbers in algebra or of distributions in mathematical analysis. These examples illustrate the ability of the language to introduce new kinds of objects, which on the ordinary understanding of the subject matter of the particular theory seem paradoxical. Nevertheless, the linguistic framework offers explicit rules for the use of these new kinds of objects and these rules enable us to reach unequivocal conclusions about several important questions. In physics the *constitutive power of language* is the ability of language to represent objects or phenomena about the existence of which we had at the earlier stages of the development of science no idea and the attributes of which seemed even contradictory.

As an illustration of the constitutive power of the language of Newtonian physics we can take the *attractive force acting at a distance between any two macroscopic bodies.* In ordinary experience we have not the slightest idea that, for example, a cup of tea is attracting my pen. And the idea that a body can act at another one at a distance seemed for many contemporaries of Newton contradictory. Newton, however, postulated the universal gravitational force and gave an exact formula expressing its magnitude. By means of this force he explained several phenomena in astronomy and mechanics.

Similarly, as Newtonian physics enriched our view of reality by the force of universal gravitation, physics of continua and fluids enriched it by *entropy*. The concept of entropy was introduced in 1865 by Rudolf Julius Emanuel Clausius in his analysis of the theory of heat engine developed by Nicolas Léonard Sadi Carnot in his *Réflexion sur la puissance motrice de feu et sur les machines propres a développer cette puissance* (Carnot 1824). Just like with gravitational attraction between macroscopic bodies, also with entropy we have no direct macroscopic experience. Nevertheless, the linguistic framework allows us to introduce this notion in an unequivocal manner.

As an illustration of the constitutive power of the language of physics of atoms energies we may take the concept of the *atom*. Matter, as it was understood by Euler was something almost mathematical. It had various attributes such as elasticity, density, color and so on, but these could be changed under external influences, as it was illustrated by Descartes on the example of wax. When we heat a piece of wax, it loses its firmness and color and receives new attributes such as transparency and liquidity. The attributes of matter are therefore not fixed; they do not determine the essence but are only accidences that can be changed. Against the background of this notion of matter the idea of atoms is something revolutionary, and again with no support in ordinary experience. It is a theoretical posit that plays a central role in the entire linguistic framework of physics of atoms and energies.

In case of the language of field theory, the radically novel by which it enriched our picture of reality is the concept of *space-time*. We experience space and time very differently and it was only the theory of relativity that revealed their deep unity.

3.7 Concluding Remarks Concerning the Potentialities of the Language of Physics

The reader may be surprised that the list of the potentialities of language does not include something like the predictive power. In the history of science, we can find many spectacular predictions, it is sufficient to mention Newton's prediction of the return of Halley's comet or Einstein's prediction of bending of the rays of light near the surface of the Sun. However, these predictions combine an *epistemological fact*, namely that language can analytically deduced the existence of the particular phenomenon, with a *contingent historical fact* that the derivation was given prior to the empirical discovery of the phenomenon. From an epistemological point of view there is no difference whether a theory predicts a phenomenon or only explains it after its empirical discovery. What is important is the existence of a link between the principles of the theory and the particular phenomenon. From a *psychological point of view*, it is a big difference — the prediction of phenomena has an air of divination — but from the epistemological point of view this difference is not significant.

When we develop an epistemological reconstruction of the development of knowledge, the time in which this development is embedded, is not the historic time of particular discoveries. Epistemology abstracts from historical contingencies and reconstructs only the interdependences of epistemic phenomena. Like Frege in his logic abandoned the Aristotelian distinction between subject and predicate and declared it only a matter of *rhetorical emphasis*, it is possible that we will have to abandon in the same radical manner the classical epistemological distinction between prediction and derivation and declare it only a matter of *psychological emphasis*.

4 The Constitution of the Potentialities of the Language of Physics

The potentialities of language, which we described in part 3 of this paper, are objective features of the language of physics. This fact is interesting in itself, because it shows that the linguistic framework determines to a certain degree the space of possibilities in which the scientific research takes place. Thus the context of discovery is a matter not only of psychology and cannot be reduced to the question of when and who proposed an idea, as the logical positivists often claimed. A discovery takes place in a linguistic framework that has some *analytic*, *expressive*, *methodical*, *integrative*, *explanatory*, and *constitutive* power and thus it determines the space, in which the psychological process of discovery takes place. A reconstruction of this space of possibilities, thus an attempt to understand what was at a given moment of history possible to discover and what not, is a matter of a philosophical and not only a psychological analysis. To delineate the limits of the possibilities opened by language of science is a philosophical problem that has remained intact, because philosophy of science was restricted to the context of justification. I have dealt with limits of the language of science on *Jornadas sobre Límites de la Ciencia* (*XVIII Jornadas de Filosofía y Metodología actual de la Ciencia*) at the University of A Coruña 14–15 March 2013 (see Kvasz 2016), and so I will not touch this topic in detail here.

But besides the philosophical analysis of the possibilities that language opens to scientific thought, the potentialities of language raise even a more serious philosophical question. It is the question of what constitutes these potentialities. In the case of mathematics, it was possible to answer this question (see Kvasz 2012) because there I could rely on Frege's reconstruction of the history of mathematics in his *Funktion und Begriff* (Frege 1891). In this paper Frege writes:

> If we look back from here over the development of arithmetic, we discern an advance from level to level. At first people did calculations with individual numbers, 1, 3, etc.

$$2+3=5\quad 2\cdot 3=6$$

are theorems of this sort. Then they went on to more general laws that hold good for all numbers. What corresponds to this in symbolism is the transition to the literal notation. A theorem of this sort is

$$(a+b)\cdot c=a\cdot c+b\cdot c.$$

At this stage they had got to the point of dealing with individual functions; but were not yet using the word, in its mathematical sense, and had not yet formed the conception of what it now stands for. The next higher level was the recognition of general laws about functions, accompanied by the coinage of the technical term „function." What corresponds to this in symbolism is the introduction of letters like *f*, *F*, to indicate functions indefinitely. A theorem of this sort is

$$\frac{dF(x)\cdot f(x)}{dx}=F(x)\cdot\frac{df(x)}{dx}+f(x)\cdot\frac{dF(x)}{dx}$$

Now at this point people had particular second-level functions, but lacked the conception of what we have called second-level functions. By forming that, we make the next step forward. (Frege 1891, 30; English translation, 40)

In this passage Frege described the type of propositions that can be demonstrated in the particular language, characterized the logical power of that language, and correlated it with the different kinds of variables. Thus, he indicated that the *logical power of language* is constituted by the *way how the language expresses generality*, or in other words by the *kind of variables that the language uses*. Led by Frege, I searched for aspects of language that are tied to the remaining potentialities in the same way as the kind of variables is tied to logical power.

It turned out that the expressive power of language is constituted by *the way how language generates complexity*. This feature can be recognized when we focus on the syntactic rules by means of which language *combines elementary symbols* and creates complex expressions. For example the language of algebra creates by constant repetition of multiplication by x (i.e. the symbol that underlies its logical power) the successive powers of the unknown x, x^2, x^3, x^4, …. (and then combines these expressions to form a polynomial). We can interpret the first power as *length*, the second as *area*, and the third as *volume*. Nevertheless, for further powers we lack any geometrical interpretation. The introduction of higher powers means thus an increase of the expressive power of language.

The paradigmatic example of the *methodical power of language in mathematics* is the analytic method developed in algebra by Viète in his *In Artem Analyticam Isagoge* (Viète 1591). This method is associated with the ability of language to express explicitly the epistemic difference between what we already know and what we try to find out in the form of the *difference between an unknown and a parameter. Parameters* (a, b, c, …) denote quantities that we know, while the unknowns (x, y, z, …) indicate quantities that we want to determine. The aspect of language that constitutes its methodical power is the *introduction of parameters*, i.e. symbols of a different epistemological but the same syntactic quality as unknowns.

I will not continue in this description; an account of the constitution of the remaining three potentialities of the language of mathematics can be found in (Kvasz 2012). I wanted only to show that thanks to Frege's analysis of the development of the language of mathematics it was possible to develop a relatively complete theory describing the constitution of the potentialities of the language of mathematics. This theory explains which aspect of the language of mathematics constitutes which particular potentiality. Unfortunately, there is nothing comparable to Frege's analysis of the development of the language of mathematics at hand for physics. The logical positivists espouse logicism as a philosophy of mathematics, and so they considered mathematics as reducible to logic. Thus they did not study the mathematical aspects of the language of physics. They focused on logical reconstruction of scientific theories, leaving the question of mathematical language out of the spotlight. So for physics, in contrast to mathematics, we do not know which aspect of language constitutes which of the particular potentialities.

Fortunately, this is not completely true, because in the case of the *methodical power of language* we have found such an aspect — it is the *new universal constant*. The example of field theory illustrates the whole situation. Before Maxwell (we have to start with Maxwell, because Maxwell's equations are Lorentz invariant, even if Maxwell did not know it), Lorentz, Poincaré and Einstein it was not clear that the speed of light is constant. The language of field theory made it possible to *stabilize the speed of light in the role of a universal constant*. After this was achieved, this new universal constant made it possible to find relations between quantities, which became the basis of new measurement methods, which led to an *increase of the methodical power of language*. Unfortunately, the methodical power of language is in this respect alone. For the other potentialities of the language of physics we do not know which particular aspects of language constitute them.

The main problem is that we do not know *what constitutes the analytic power of the language of physics*. In mathematics, as Frege has found, it is the way how language expresses generality, i.e. the *kind of variables* by means of which generality is expressed. Mathematics is a discipline that studies relations of logical consequence among propositions and the variables are instruments that make it possible to transfer generality of a particular kind from the assumptions of a judgment onto its conclusion. Physics instead of logical consequence (which is *atemporal*) examines the relations of causal consequence (expressed by means of differential equations describing the *temporal* evolution of the state). It is not clear what can play in the case of a differential equation a role analogous to the role of variables in a logical inference. We would need a philosophical analysis of the language of differential equations used in physics comparable to Frege's analysis of the linguistic tools used in mathematics to express generality. I believe that this is an interesting and important philosophical question, which unfortunately was not yet raised at all.

And of course, by analogy we can ask *what constitutes the other potentialities of the language of physics*. When we succeeded in answering, at least partially, these questions, I believe that we would achieve a substantial progress in understanding the role of language in scientific research. The methodical power indicates that this is not a completely hopeless task.

Acknowledgment The author expresses his acknowledgement of the generous support of the project *Formal Epistemology — the Future Synthesis,* in the framework of the program *Praemium Academicum* of the Czech Academy of Sciences.

References

Bohr, N. (1913). On the constitution of atoms and molecules. *Philosophical Magazine, 26*, 1–25.

Carnot, S. (1824). *Réflexion sur la puissance motrice de feu et sur les machines propres a développer cette puissance.* Paris: Chez Bachelier.

Einstein, A. (1905a). Zur Elektrodynamik bewegter Körper. *Annalen der Physik, 17*, 891–921.

Einstein, A. (1905b). Über einen die Erzeugung und Verwandlung des Lichtes betreffenden heuristischen Gesichtspunkt. *Annalen der Physik, 17*, 132–148.

Euler, L. (1748). Sur la vibration des cordes. *Mémoires de l'Academie Royale des Sciences et des Belles-Lettres de Berlin, 4*, 69–85.

Euler, L. (1750). Découverte d'un nouveau principe de mécanique. *Mémoires de l'Academie Royale des Sciences et des Belles-Lettres de Berlin,* 6, 1750, 1752, 185–217.

Euler, L. (1755). Principes géneraux du movement des fluides. *Mémoires de l'Academie Royale des Sciences et des Belles-Lettres de Berlin, 11*, 274–315.

Faraday, M. (1831). *Experimental Researches in Electricity.* Reprinted in M. Hutchings (ed. 1955). *Great Books of the Western World.* London: Encyclopedia Britannica Inc.

Fourier, J. (1822). *Théorie Analytique de la Chaleur.* Paris. Translated by A. Freeman as (1955) *The analytical theory of heat.* New York: Dover.

Frege, G. (1891). Funktion und Begriff. Reprinted in Frege, G. (1989), *Funktion, Begriff, Bedeutung.* Göttingen, Vandenhoec & Ruprecht, (pp. 17–39). English translation in: P. T. Geach & M. Black (Eds.) 1952). *Translations from the philosophical writings of Gottlob Frege* (pp. 21–42). Oxford: Basil Blackwell.

Galileo (1610). *The Starry Messenger.* In S. Drake (1957), *Discoveries and opinions of Galileo* (pp. 229–280). New York: Doubleday Co.

Kant, I. (1755). *Universal natural history and theory of the heavens.* Translated by W. Hastie (1969), Ann Arbor: University of Michigan Press.

Kvasz, L. (2000). Changes of language in the development of mathematics. *Philosophia Mathematica, 8*, 47–83.

Kvasz, L. (2012). *Language in change. Fernando Gil International Prize 2010.* Lisbon: Fundacao Calouste Gulbenkian.
Kvasz, L. (2016). Language and the limits of science. In W. J. Gonzalez (Ed.), *The limits of science. An analysis from "barriers" to "confines"* (pp. 69–93). Leiden: Brill Rodopi.
Laplace, P. S. (1796). *Exposition du systéme du monde.* Paris: De l'imprimerie du Cercle-Social.
Laplace, P. S. (1816). Sur la vitesse du son dans l'air et dans l'eau. *Annales de chimie et de physique, 3,* 238–241.
Lavoisier, A. L. (1789). *Traité elémentaire de chimie.* Paris. English translation by R. Kerr (1952), *Elements of chemistry. Great books of the western world,* vol. 45. Chicago: Encyclopedia Britannica.
Maxwell, J. C. (1861). On the physical lines of force. *Philosophical Magazine, 21,* 161–175.
Maxwell, J. C. (1873). *A treatise on electricity and magnetism.* Oxford: Clarendon Press.
Navier, H. (1822). *Mémoire* sur le *lois du mouvement des fluids. Mémoires de l'Acadèmie royale des sciences,* pp. 389–440.
von Neumann, J. (1927). Mathematische Begründung der Quantenmechanik. *Nachrichten von der Gesellschaft der Wissenschaften zu Göttingen, Mathematisch-Physikalische Klasse,* pp. 1–57.
Newton, I. (1687). *The principia.* A new translation by I. Bernard Cohen & A. Whitman (1999), Proceeded by *a guide to Newton's principia.* Berkeley: University of California Press.
Planck, M. (1900). Zur Theorie des Gesetzes der Energieverteilung im Normalspektrum. *Verhandlungen der Deutschen physikalischen Gesellschaft, 2,* 237–245.
Poisson, S. D. (1823). Sur la vitesse du son. *Annales de chimie et de physique, 23,* 5–16.
Taylor, B. (1713). De motu nervi tensi. *Philos. Transactions of the Royal Society, 28,* 26–32.
Viète, F. (1591). *In Artem Analyticam Isagoge.* Translated by T. R. Witmer (1983) as *Introduction to the analytical art.* Kent: The Kent State University Press.

Part II

Language and Change in Scientific Research: Evolution and Historicity

4

Scientific Inquiry and the Evolution of Language

Jeffrey Barrett

1 The Coevolution of Theory and Language

Having an appropriately rich descriptive language is a precondition for the practice of empirical science. But descriptive language is also a *product* of empirical inquiry. When we characterize an event as occurring in a particular region of *curved spacetime* or describe an *electron* as being in a *eigenstate* of *z-spin*, and hence in a *superposition* of being *x-spin up* and *x-spin down*, we are using language that evolved in the context of empirical inquiry to address the descriptive and explanatory demands we in fact faced in better understanding nature.

Our descriptions of the physical world are the result of empirical exploration. Our best and most precise descriptive language has been forged in the context of ongoing empirical inquiry for the sake of successful prediction, explanation, and action. The more precise and rigorous

J. Barrett (✉)
Department of Logic and Philosophy of Science, School of Social Sciences,
University of California at Irvine, Irvine, CA, USA
e-mail: j.barrett@uci.edu

W. J. Gonzalez (ed.), *Language and Scientific Research*,
https://doi.org/10.1007/978-3-030-60537-7_4

our predictive and explanatory demands, the more sophisticated our evolved descriptive language and its associated concepts must be.

Inquiry does not involve simply choosing a particular descriptive language, specifying a set of hypotheses in that language, then updating one's degrees of belief in the various hypotheses under consideration as one gets on new empirical evidence. Rather, our beliefs about the physical world and the language we use to express these beliefs *coevolve.* Our best understanding of the world is a function of the descriptive resources that have coevolved to facilitate and represent that understanding. That our best theories and the descriptive language and concepts associated with those theories manifestly coevolve demands a more subtle account of inquiry than what one gets from a standard belief-revision model of knowledge with a fixed descriptive language.

W. V. Quine sought to address this issue by means of a belief-revision account of empirical knowledge where one's conceptual scheme might itself evolve over time. To this end, Quine characterized our beliefs concerning the nature of the world as a system that is only *in part* constrained by experience. As he put the point, "[t]he totality of our so-called knowledge or beliefs, from the most casual matters of geography and history to the profoundest laws of atomic physics or even of pure mathematics and logic, is a man-made fabric which impinges on experience only along the edges" (1980, 42).[1] Because our beliefs and commitments work together to account for experience, "it is misleading to speak of the empirical content of an individual statement" (1980, 43). Rather, "[e]ach man is given a scientific heritage plus a continuing barrage of sensory stimulation; and the considerations which guide him in warping his scientific heritage to fit his continuing sensory promptings are, where rational, pragmatic" (1980, 46). The resulting picture is one where meanings shift as one reevaluates the truth conditions of one's statements. As a result, the semantic content of the language one uses to describe the world coevolves as one revises one's theoretical commitments to fit experience.

[1] All of the quotations from Quine here are from the last section of his 1951 paper "Two Dogmas of Empiricism." It is here where he gives his positive view of nature of empirical inquiry in terms of a pragmatic belief-revision model of knowledge.

Quine's methodological suggestion was that one should revise both one's beliefs and the language one uses to express those beliefs with the aim of "working a manageable structure into the flux of experience" (1980, 44). But, even so, he held that how one revises one's theoretical and linguistic commitments is radically underdetermined by experience and such pragmatic advice. Because our best understanding of the world only makes contact with experience along the edges, by means of our most empirically descriptive language, there will always be multiple ways to square one's theoretical and linguistic commitments with recalcitrant experience. Among other options, one might end up revising one's predictive theory or one's descriptive language or both.

In a similar spirit, Thomas Kuhn famously noted that there was significant historical reason to believe that our descriptive language's meaning in fact coevolves with our best theories.[2] Indeed, even a cursory consideration of the history of empirical inquiry provides compelling evidence that our theories and descriptive language coevolve. Our descriptive language evolves by *semantic drift* (whales are not *fish* after all), *invention* (the two particles in an EPR state are *quantum-mechanically entangled*), and *discard* (planets do not move on *epicycles*). This evolution is theory driven, and how theory and language coevolve is indeed underdetermined by our basic empirical experience and pragmatic considerations.[3]

For his part, Kuhn was keen to emphasize the problems that the coevolution of theory and language pose for the tradition view of empirical science as progressing toward a true description of the world. In describing how competing paradigms are incommensurable, he reported that

> Successive paradigms tell us different things about the population of the universe and about that population's behavior. They differ, that is, about such questions as the existence of subatomic particles, the materiality of light, and the conservation of heat or of energy. (Kuhn 1996, 103)

---

[2] The first edition of Kuhn's *Structure of Scientific Revolutions* was published in 1962, a few years following Quine's "Two Dogmas of Empiricism."

[3] For just a bit more detail on the three examples, *epicycles* are no longer required to explain the retrograde motion of planets, the *entangled* states of particles at difference locations explain the empirical violation of Bell-type inequalities, and we needed some biological theory to properly classify whales.

Because of the radical difference in how the language associated with competing paradigms evolves, he argued that "there can be no scientifically or empirically neutral system of language or concepts" (1996, 146). And that, hence, one can never neutrally compare the scientific virtues of alternative theories and their associated descriptions of the natural world.

Such considerations led Kuhn to conclude that "[w]e may …have to relinquish the notion, explicit or implicit, that changes of paradigm carry scientists and those who learn from them closer and closer to the truth" (1996, 170). The picture he presented was one where scientific knowledge has progressed "as we now suppose that biological evolution did, without benefit of a set goal, a permanent fixed scientific truth, of which each state in the development of scientific knowledge is a better exemplar" (1996, 172–3). Kuhn took the fact that our best theories, descriptive language, and explanatory and predictive aims coevolve to undermine the view that science is progressing toward a true description of the world.

But this goes too far. That our best theories and descriptive language result from a contingent evolutionary process does not entail that they fail to track objective facts. While biological evolution is a deeply contingent process with no specifiable ultimate goals, it is nevertheless constrained by facts regarding the relative fitnesses of alternative adaptions. In this sense, an ongoing evolutionary process, while in some ways aimless, will work to tune the evolved capacities and behaviors of the resultant organisms to those objective truths that are in fact relevant to their fitness.

By analogy, one might consider what facts one should expect to constrain the evolution of our best descriptive language and theory in inquiry. While there is nothing to *guarantee* convergence to the truth, one might well expect to eliminate significant descriptive error insofar as inquiry answers to one's actual success and failure in action and descriptive error often stands in the way of successful action. While it is not immediately clear the precise sense in which this might count as progress toward *truth*, there is good reason to believe that the coevolution of theory and descriptive language is at least not incompatible with such progress. Indeed, we will see that one can say rather more in defense of scientific progress toward the truth on an evolutionary model.[4]

An adequate account of scientific knowledge must be capable of modeling the coevolution of our best empirical theories and the associated descriptive language. Here we will use the tools of evolutionary game theory to illustrate how a simple model for the coevolution of predictive theory and language might work. The model will show how it is possible for incommensurable languages to evolve in the context of changing predictive and descriptive demands. It will also show how such incommensurability may nevertheless be compatible with a commitment to science providing true descriptions.

2 A One-Sender/One-Predictor Game

There are two basic types of evolutionary models. In a *learning model* one describes how the dispositions of a set of agents changes as they learn from their interactions with each other and the world. In a *population model* one describes how the proportion of agents of different types in a population changes as traits are passed from one generation to the next given environmental constraints. One might also consider hybrid evolutionary models where agents learn, then pass on their inheritable traits, then learn again. Importantly, a particular learning model often naturally corresponds to a particular population model. Other things being equal, for example, the dispositions of agents who learn by simple reinforcement can be expected to track the distribution of types of agents in populations that evolve under the replicator dynamics.[5] For the sake of simplicity and uniformity, we will consider learning models here and in the companion paper. Similar evolutionary stories can be told in the context of population models where successful traits are passed from one generation to the next.

[4] And we will consider the evolution of the notion of truth itself in the companion paper (Barrett 2021) in this volume.

[5] See Skyrms (2006, 2010) and Argiento et al. (2009) discussions of reinforcement learning. See Skyrms (2010) for an account of how to translate results from a learning model to a population model by considering the mean field dynamics.

The learning models we will consider are generalized Lewis-Skyrms signaling games.[6] We will start with a simple example to show how such games work. We will then consider slightly more subtle signaling games and use them to model correspondingly more subtle features of empirical inquiry.

The first model is something that one might think of as a sender-predictor game. The basic structure of the game is represented in the figure below (Fig. 4.1).

On each play of the game, the sender observes the state of nature in the morning, then sends a signal to the receiver, who does not have access to the state of nature. The receiver performs a predictive action for the afternoon that is either successful or unsuccessful. We will suppose here that if the act is successful, then the disposition that led to each agent's last action is reinforced and that otherwise it is weakened. More specifically, we will suppose that the agents learn by *bounded reinforcement with punishment*. We will also suppose that the laws of nature that tie afternoon states to morning states are deterministic.[7]

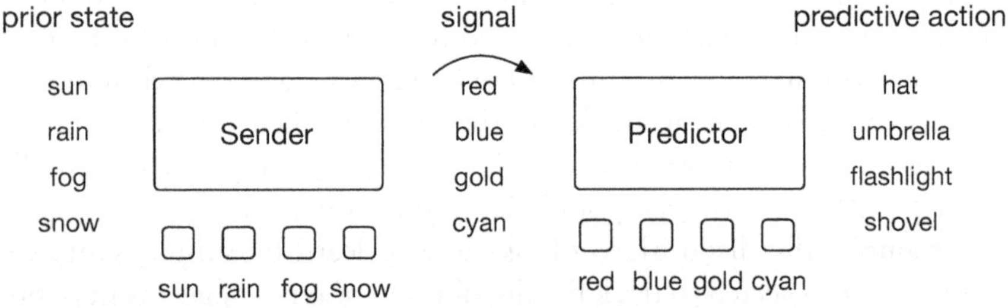

Fig. 4.1 A one-sender/one-predictor game (Source: Author)

[6] Lewis (1969) first presented this sort of game in the context of classical game theory as a way of studying how conventions might be established. Skyrms (2006, 2010) translated Lewis' signaling games into the context of evolutionary game theory. Here one need not suppose that the agents are in any way rationally sophisticated. Rather, in an evolutionary game, the agents may start with no knowledge whatsoever and gradually learn as they interact with each other and the world. Further, there is no need for some sort of natural salience to break the symmetry between equally good conventions. In an evolutionary game, such symmetries may be broken by random fluctuations in the dispositions of the agents as they evolve.

[7] We will consider other learning dynamics in other models. Different learning dynamics exhibit different properties. Indeed, as we will see in the companion paper, the agents in the present game

One might imagine each agent as having a collection of urns each containing a number of balls. The sender has four urns labeled *sun*, *rain*, *fog*, and *snow*. Each of these urns starts with one ball of each color *red*, *blue*, *gold*, and *cyan*. The receiver, who will make a prediction based on the signal he receives, has four urns labeled *red*, *blue*, *gold*, and *cyan*. Each of these urns starts with one ball each labeled *hat*, *umbrella*, *flashlight*, and *shovel*.

The inquirers' first-order dispositions to signal and act are determined by the proportional contents of their urns. Their second-order dispositions to learn from the success and failure in action are determined by how the contents of their urns are updated given their experience.

The sender observes the weather in the morning, draws a ball from the corresponding urn, then sends a signal by waving a flag of the color corresponding to the ball she draws. The receiver sees the signal, draws a ball from the urn corresponding to that color, then performs the corresponding predictive action that afternoon. Supposing deterministic laws we will suppose that the afternoon action of bringing a *hat*, *umbrella*, *flashlight*, or *shovel* to the picnic is successful if and only if the morning state was *sun*, *rain*, *fog*, and *snow* respectively.

We are supposing that the agents learn by bounded reinforcement with punishment. Here this works as follows. If the receiver's predictive action was successful, each agent returns the ball she drew to the urn from which it was drawn, then adds a ball of the same type to that urn unless there are already 1000 balls of that type in the urn. Otherwise, each agent discards the ball she drew unless it was the last ball of its type in the urn from which it was drawn, in which case she simply returns the ball to the urn from which it was drawn.

On simulation, when the modeled sender and receiver update their signaling and predicting dispositions using this dynamics, they start off randomly signaling and randomly acting. The signals initially have no meanings and there is no pattern to the predictive dispositions of the receiver. But, as they learn from experience, they typically (0.993) evolve

do not do nearly as well evolving an optimal descriptive language and predictive dispositions if they learn by simple reinforcement without bounds or punishment.

a set of nearly optimal (0.994) descriptive and predictive dispositions.[8] Here the modeled inquirers nearly always coevolve successful, systematically interrelated, descriptive and predictive dispositions.[9]

In order to be successful, the inquirers must *coevolve* the ability to faithfully represent prior morning states and to reliably predict posterior afternoon states. When they successfully evolve an adequate descriptive language and predictive dispositions, their first-order dispositions to signal and act evolve to be stable under their second-order dispositions to learn. The inquirers' first-order dispositions evolve to represent a sort of practical knowledge that allows them to satisfy the norms represented by their second-order dispositions to update their first-order dispositions.

3 A Two-Sender/One-Predictor Game

Evolutionary games that involve more subtle interactions between inquirers may lead to correspondingly more subtle evolved dispositions. Consider a game with two senders and one receiver who makes predictions based on information for each sender as represented in the following figure (Fig. 4.2).[10]

Both senders observe the full state of nature, then send their signals. The receiver, who know which signal was sent and who sent it, then performs a predictive action. If the act is successful, then the disposition that led to each agent's last action is reinforced; otherwise it is weakened. We will again assume bounded reinforcement learning with punishment and deterministic natural laws connecting morning states of nature to afternoon states.

Again, one might imagine each agent as having a collection of urns each containing a number of balls. Here each sender has four urns labeled 0, 1, 2, and 3 corresponding to the four morning states the sender might observe. Each of these urns starts with one 0 ball and one 1 ball

[8] These are the mean cumulative success rates on 10^3 runs with 10^6 plays per run.

[9] As we will see in the companion paper, the level of success in this game is a function of the particular learning dynamics.

[10] This sort of dynamic partitioning game was first presented in Barrett (2006) and discussed in Barrett (2007a and 2007b).

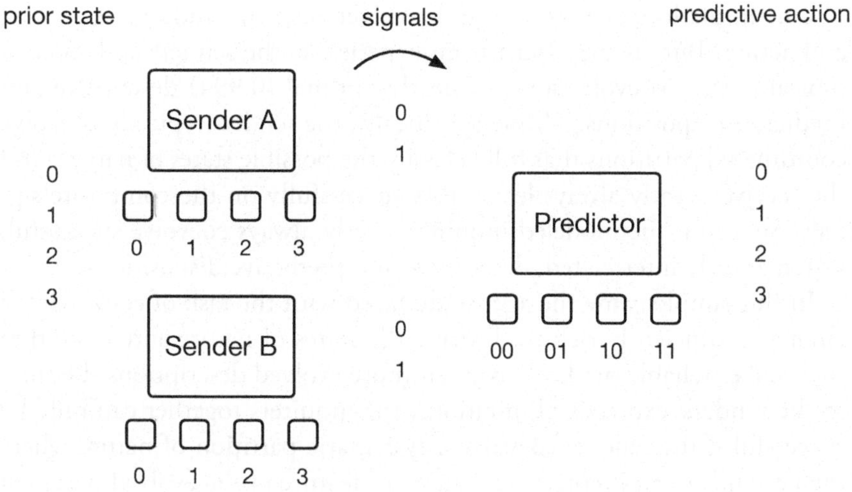

Fig. 4.2 A two-sender/one-predictor game (Source: Author)

corresponding to signal types. The receiver has four urns labeled 00, 01, 10, and 11 corresponding to each possible ordered pair of signal she might receive from the two senders (the first term indicates the signal from sender *A* and the second from sender *B*). Each of these urns starts with one ball labeled 0, 1, 2, and 3 corresponding to the predictive actions the receiver might perform in the afternoon. We will suppose that the predictive action is successful if and only if it matches the morning state.

So on this model, just as on the last, there are four states of nature and four possible predictive actions that the predictor might perform, The difference is that here there are two senders observing the world and, at least initially, sending two completely independent signals. Since they only have two signal types, neither sender has the linguistic resources to characterize the full state of nature. Consequently, the only way the inquirers can be successful is if the senders coevolve systematically interrelated signals and the receiver coevolves the ability to interpret the composite signals for the purpose of successful predictive action.

On simulation, the senders and receiver start by randomly signaling and acting. But, as they learn from experience, the senders and receiver typically (0.995) evolve a set of nearly optimal (0.994) descriptive and predictive dispositions.[11] More specifically, the senders typically coevolve coordinated partitions that fully classify the possible states of nature, and the receiver nearly always learns acts successfully on the composite signals. So again, the modeled inquirers nearly always coevolve successful, systematically interrelated, descriptive and predictive dispositions.

In this simple game the agents are faced with the task of coevolving a language sufficiently rich to describe the states of nature and a law that that makes reliable predictions given those evolved descriptions. Because of the senders' expressive limitations, the inquirers together can only be successful if the senders coevolve a systematic partition of nature where each attends to and represents a different feature of the evolved partition and the receiver coevolves a law that allows for successful predictions given the senders' representations. The features the senders attend to, are not preestablished natural kinds that "carve nature at it joints" to use Quine's expression. Rather, they are simply coordinated representations that have evolved to facilitate successful predictive action. The inquirers might have evolved a different partition of nature. Indeed, they do on different runs of this game. But, given the success of the terms they use to describe nature and make predictions, *from the perspective of the inquirers* their evolved terms might well *appear to pick out natural kinds.*[12]

4 Revolution in the Context of Changing Demands

In the two games we have discussed so far, the inquirers coevolve interrelated descriptive and predictive dispositions. The way that these dispositions are intertwined can be better understood by considering what

[11] This is again on 10^3 runs with 10^6 plays per run.

[12] See Barrett (2007a) for a discussion of the evolution of natural kind language.

happens when the laws of nature or the second-order dispositions of the agents change during inquiry. To this end, we will consider a slightly more subtle game.[13]

The new game is precisely the same as the last except that the inquirers' second-order dispositions to update their descriptive and predictive dispositions change after they have evolved an initial set of successful descriptive and predictive dispositions. More specifically, after the agents have evolved successful first-order descriptive and predictive dispositions, as we have seen they nearly always do, we will suppose that their second-order dispositions change in such a way that they begin to reinforce action 2 on state 1 and action 1 on state 2. Since their second-order dispositions determine what in fact counts as successful actions that get reinforced or unsuccessful actions that get punished, one might think of changing the agents' second-order dispositions as changing their values or utilities.

Importantly, changing how the agents reinforce in this particular model may be understood *equally well* as changing their preferences or as changing the natural laws that govern the world that they inhabit. In particular, the change from the first-stage to the second state of the game might be understood as one where nature has adopted a new set of natural laws that connect morning states to afternoon states. The new laws are such that the afternoon state that used to occur when the morning state was 1 is now the afternoon state that used to occur when the morning state was 2 and the afternoon state that used to occur when the morning state was 2 is now the afternoon state that used to occur when the morning state was 1 (Fig. 4.3).

The questions for the new model are (1) whether the inquirers are able to adapt their previously evolved first-order dispositions to their new second-order dispositions to reinforce and (2) if they are, how do they adapt to the new pragmatic demands of inquiry. In answer to the first question, the senders and receiver typically (0.972) do evolve a second set of nearly optimal (0.993) linguistic and predictive dispositions with the

[13] This is a variation of the two-sender game above. This type of game was suggested in conversation by Michael Dickson. It is first discussed in Barrett (2009).

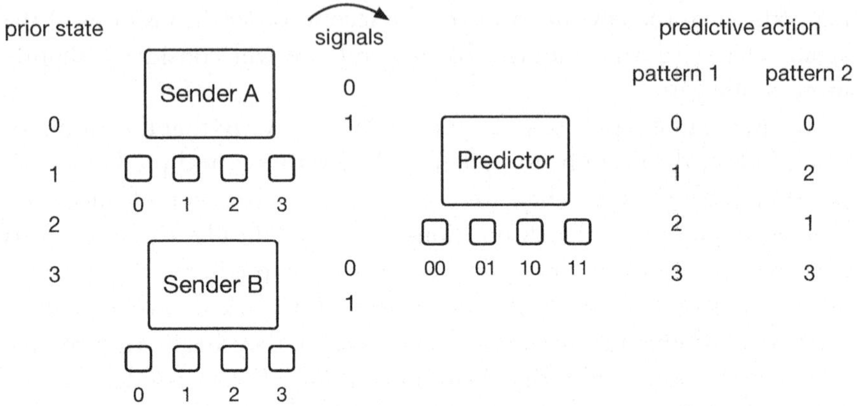

Fig. 4.3 A sender-predictor game with changing demands (Source: Author)

new reinforcements in the second stage of the game.[14] But the answer to second question is a bit more subtle. Just as Quine had it, it is underdetermined precisely *how* the agents will evolve the new set of dispositions that is adapted to the new demands. Sometimes the *receiver* evolves to make predictions differently (0.581), sometimes the two *senders* evolve new coordinated partitions that classify the states of nature differently (0.419), and for repeated changes in the second-order constraints the agents may employ a combination of both of these strategies.

When the receiver's *predictive dispositions* are revised, the senders associate the same signal with each state of nature, but the receiver changes how he makes predictions on the signals. An example of this is illustrated in Fig. 4.4. Here the senders associate the same signal with each state of nature, but the receiver changes how he makes predictions on the signals. The way that the inquirers' terms partition nature stays the same, but they evolve a new predictive law.

In contrast, when the sender's *descriptive dispositions* are revised, they associate new signals with the states of nature, and the receiver makes

[14] The second stage of the game is run with the same parameters as the first stage.

old language and predictive theory — stage 1

prior state	A sends	B sends	receiver does
0	1	0	0
1	0	1	1
2	1	1	2
3	0	0	3

new standard of success

new language and predictive theory — stage 2

prior state	A sends	B sends	receiver does
0	1	0	0
1	0	1	2
2	1	1	1
3	0	0	3

Fig. 4.4 Sometimes the receiver changes her predictive dispositions (Source: Author)

predictions on the signals precisely the same way as before. An example of this is illustrated in Fig. 4.5. Here the senders associate new signals with the states of nature, and the receiver makes predictions on the signals exactly the same way as before. The inquirers' terms evolve to partition nature differently, and it may well appear to them that they have discovered a new set of natural kinds.

As predictive demands change repeatedly over time, the modeled inquirers' descriptive and predictive dispositions *coevolve* and one sees a mixture of both of these phenomena. Sometimes the inquirers evolve a new predictive law, sometimes they evolve a new language that carves nature differently, and sometimes they do a bit of each.

That the modeled agents change their predictive dispositions more often than they change their language to accommodate their new old language and predictive theory — stage 1

prior state	A sends	B sends	receiver does
0	0	1	0
1	1	0	1
2	0	0	2
3	1	1	3

new standard of success

new language and predictive theory — stage 2

prior state	A sends	B sends	receiver does
0	0	1	0
1	0	0	2
2	1	0	1
3	1	1	3

Fig. 4.5 Sometimes the senders evolve new linguistic dispositions (Source: Author)

predictive demands is perhaps unsurprising. Given that each sender only has two terms and that they hence need to coordinate how they use their terms, it is relatively difficult for them to evolve a language that systematically partitions nature and allows them to individuate the four salient states. For this reason, evolving a new language from scratch to address new predictive demands is nontrivial. On the other hand, if the inquirers have already evolved a perfectly expressive language, then that language, whatever it is, can still be used to let the receiver know the morning state. He just needs to learn to act differently on this information. And that is precisely what happens here more often than not.

Given the difficulty of evolving a descriptive language from scratch, one might expect language to be more stable than theory more generally. Similarly, one might further expect that the more complicated and successful the initially evolved language, the less likely that it will undergo a radical change in the context of new predictive demands. Parts of it may end up being refashioned to the new context, but one would expect that the evolution of new predictive actions based on the same descriptions to occur more often, just as illustrated in the present model. Of course, the only way to say what would in fact happen on a particular game is to see. As the present case illustrates, even very simple models may exhibit surprisingly subtle behavior.[15]

The inquirers in the present game are typically able to evolve a new optimal language because the learning dynamics involves *bounded reinforcement* and allows for *punishment.* This means that whatever dispositions are initially learned can be *unlearned* in the context of the new predictive demands.[16]

Of course, the senders sometimes do evolve a new descriptive language here. When they do, their old and new languages are typically *incommensurable* in the sense that they are not term-by-term intertranslatable. This is a result of the evolved terms in the two languages corresponding to fundamentally different ways of carving up nature. The figure below shows the relationship between sequentially evolved languages on an actually run of the game. While each language allows the agents to perfectly individuate the states of nature and, hence, allows for perfectly successful action, the two languages are term-wise incommensurable (Fig. 4.6).

	B0	B1
A0	2	0
A1	1	3

	B0	B1
A0	1	0
A1	2	3

Fig. 4.6 The incommensurability of sequentially evolved terms (Source: Author)

[15] To get a sense of how the behavior of this present model scales, one might try an eight-state, three-sender (each with only two terms) game with changing reinforcements. One would predict on the present considerations that the receiver would revise his predictive behavior yet more often in the context of the more complicated evolved language.

[16] See Barrett and Zollman (2009) for a discussion of the role of forgetfulness in learning.

Here B1 means the same thing in both languages in terms of the states of nature, but no term in the old language translates the new term A0. While the two languages are term-wise incommensurable, this incommensurability does not represent any failure in the agents' ability to faithfully describe their world. This can be seen from the modeler's god's-eye-view. The agents need to coevolve the ability to individuate the four states of nature and make different predictions in each to be maximally successful here, and they do precisely that in each language.

While the two sequentially evolved languages here are term-wise incommensurable, they allow the agents to individuate precisely the same states. But sometimes the type of incommensurability exhibited between the sequentially evolved languages is more complete. One way this can happen is when the evolved languages are suboptimal and do not allow the agents to individuate each possible state of nature.

Simple reinforcement learning is a learning dynamics that involves the same sort of urn learning on success that we have been considering, but balls are never removed on failure and there is no bound on the maximum number of balls of each type in an urn.[17] On this simpler learning dynamics, the present game sometimes (0.269) evolves a language that is less tightly connected to the world and consequently does not allow the agents to perfectly individuate between states of nature. It may happen, for example, that one of the senders' composite expressions is used to describe two different states of nature. While an evolved language that exhibits this sort of vagueness limits the inquirers' overall level of success, they always evolve to be more successful than not. The degree of predictive success they achieve even when they evolve a suboptimal language means that they are still getting *something* descriptively right. While the inquirers themselves will not know precisely what their representations are getting right, something we will return to in a moment, their descriptions always do in fact evolve to more-or-less accurately represent their world on this game, and this is reflected in the predictive success of the

[17] We will discuss simple reinforcement learning in more detail in the companion paper in this volume (Barrett 2021).

receiver. Indeed, the descriptive faithfulness of the evolved language *explains* the agents' success from the god's-eye-view.

One can more clearly see how this sort of incommensurability works by considering a $3 \times 3 \times 3$ signaling game with one sender and one receiver. Under simple reinforcement learning, this game sometimes (0.096) evolves to a suboptimal partial pooling equilibrium like the one illustrated in Fig. 4.7. Here the sender sends signal 0 with probability p and 1 with probability $1 - p$ in state A and sends signal 2 in state B and state C. The receiver does act A on signals 0 and 1, but on signal 2 does act B with probability q and act C with probability $1 - q$. Here p and q may be any probabilities between zero and one. On such mixed strategies, the receiver's act matches the state of nature 2/3 of the time on average. While not optimal, the sender's and receiver's evolved strategies are a Nash equilibrium since neither agent can improve their payoffs by unilaterally changing the strategy they are playing conditional on a given state of nature or signal. Alternative suboptimal languages of this type may be descriptively faithful in quite different ways, and there may hence be no precise translation at all between evolved terms or even full expressions.

A similar phenomena also occurs in signaling games where there is an information bottleneck resulting from the agents lacking sufficient linguistic resources to individuate all of the states that in fact matter for successful action. As a simple example, consider a signaling game with one sender and one receiver that has ten states of nature, two terms, and

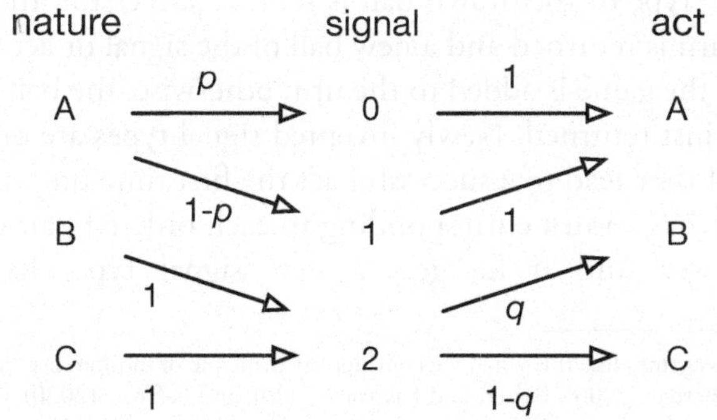

Fig. 4.7 A suboptimal 3×3×3 language with mixed strategies (Source: Author)

ten possible predictive actions. Here the two terms may end up exhibiting quite different meanings from one run of the game to the next. But the sender's descriptions may nevertheless come to describe nature as faithfully as possible given her descriptive resources. While she does not have the ability to individuate each state, she will typically evolve a partition and use each of her terms to represent one of the elements of the partition.[18] And one would expect yet more subtle sorts of incommensurability to be exhibited by models that allow for the evolution of more complicated sorts of language.

5 Invention and Discard

The two-sender/one-receiver model we have been considering illustrates how even dramatic *semantic drift* might occur under the changing demands of inquiry. One can also investigate linguistic *invention* and *discard* by means of a model with a slightly more sophisticated learning dynamics.

Suppose that the inquirers in the two-sender/one-receiver game learn by *simple reinforcement with invention.*[19] On this dynamics, each sender has an urn corresponding to each possible state of nature and each urn begins with just a single black ball. When presented with a state, each sender draws a ball at random from the corresponding urn. If the ball is black, a new signal type is invented and sent to the receiver; otherwise, a signal of the type of the drawn ball is sent. If successful, the ball drawn from each urn is returned and a new ball of the signal or act type used in that play of the game is added to the urn; otherwise, the ball drawn from each urn is just returned. Newly invented signal types are only kept and reinforced if they lead to a successful act the first time they are used. The receiver also has an urn corresponding to each ordered pair of signals he might receive. And if he gets a new signal type, he introduces

[18] We will discuss games like this when we consider the principle of indifference in the companion paper. See also Barrett (2006), Barrett and LaCroix (2020), and LaCroix (2020) a studies of information transfer in the context of such games.

[19] See Skyrms (2010) and Alexander et al. (2012) for descriptions of reinforcement learning with invention and its properties.

corresponding urns to represent the new possible pairs. Each of the receiver's urns begins with a single ball of the four act types: 0, 1, 2 and 3. Here the evolutionary game itself evolves as the linguistic resources of the agents change over time.

In the context of a two-stage game where the demands of inquiry change between stages as in the last section, agents who learn by reinforcement with invention sometimes appropriate the terms of their old descriptive language to the new descriptive and predictive demands, sometimes they invent new terms for new purposes and just stop using some of the old terms, and sometimes they pursue a combination of the two strategies. In this way, the descriptive languages of agents learning by simple reinforcement with invention may exhibit semantic drift, invention, and discard. The particular combination depends on the tasks at hand, the descriptive resources the agents have, the relative costs of using those resources, and chance events in the evolutionary process that forges their descriptive and predictive dispositions.

6 Sender-Predictor Games as Models for the Coevolution of Predictive Theory and Descriptive Language

The two-sender/one receiver game we have been discussing provides a simple model that explains how descriptive and predictive dispositions (1) may be *intimately intertwined* in empirical inquiry, (2) may *coevolve* to allow for successful action, and (3) may *continue to coevolve* to accommodate shifting descriptive and predictive demands. It further illustrates how incommensurable languages may sequentially evolve and how agents might satisfy the changing demands of inquiry by revising their descriptive *and/or* predictive dispositions. This last point provides a sharp example of the scope of empirical underdetermination.

Language and predictive practice are *co-emergent phenomena* in this model. The agents initially lack a meaningful descriptive language and their predictive practice is wildly unreliably, but they evolve acquire successful descriptive and predictive practices as their first-order dispositions accommodate to the demands of nature and the second-order dispositions that determine their payoffs.

The present model illustrates how one's descriptive theory and language coevolve in the context of changing predictive and explanatory demands, but it also shows why such an evolutionary process need not undermine the inquirers' progress toward in an increasingly faithful description of the world. While sequentially evolved languages may be incommensurable in subtle ways, they may nevertheless allow for more or less faithful descriptions of the world. While the inquirers themselves may not know precisely how their descriptions get the world right, the fact that they facilitate successful coordinated action and prediction is evidence that they are getting something right.

That expressive incommensurability is not evidence of failure in description is easily argued in the present model. It sometimes happens under simple reinforcement learning that the inquirers in the two-sender/one receiver game evolve a suboptimal descriptive language, the sort of imprecise language corresponding to a partial pooling equilibrium involving mixed strategies, then, when faced with different descriptive demands, they subsequently evolve an optimal descriptive language that allows them to perfectly individuate the states of nature. In this case, the sequentially languages will exhibit an especially pernicious sort of incommensurability. But such incommensurability need in no way undermine progress toward a language that allows for the perfectly faithful representation of nature. Indeed, in the present example the second language is perfectly suited to describing the world the inquirers inhabit given what in fact matters for successful action.

There is also a moral here for the *pessimistic meta-induction*, the argument that since our past successful scientific theories were found to be false, we have good inductive reason to believe that our currently successful theories are also false. While the descriptive theories resulting from real empirical science rarely if ever provide entirely satisfactory characterizations of the world, this does not mean that they fail to faithfully represent nature. Further, that scientific inquiry has sometimes involved abandoning one description of nature in favor of another, that does not mean that the abandoned descriptions failed to faithfully represent nature. The present model illustrates how even radical descriptive change is perfectly compatible with *both* the inquirers' descriptions of the world before the change *and* their descriptions after the change being faithful in their way. It also concretely illustrates how evolved descriptions, even in radically incommensurable languages, can all be at least approximately true. The descriptions the inquirers evolve might be accurate and perfectly suitable given their initial aims but exhibit significant and subtle changes in the context of new explanatory and predictive demands. From a god's-eye-view, we can see that the sequentially evolved descriptions that allow for successful action all allow for faithful descriptions of the world in their way, and more than one alternative descriptive framework may even be optimal given the descriptive and predictive tasks that the inquirers face.

Just as with the modeled inquirers, while our current best theories describe the world in a way that is more or less suitable in the context of our current descriptive and predictive demands, it is possible, indeed likely, that we will need to reconfigure our descriptions in the context of future explanatory and predictive demands. But this does not mean that our current descriptions of the world are not faithful. The faithfulness of our descriptions, just as with the modeled inquirers, is reflected in the success we achieve when we use them. Of course, that success does not tell us precisely what we are getting right. To know that would require that one have a god's-eye-view on the process of inquiry. But not knowing precisely what we are getting right is perfectly compatible with the manifest fact the success in action facilitated by our descriptive practice is the result of our descriptions getting *something* right, something that would explain the successful action it allows from the god's-eye-view.

Such abstract philosophical points can be made clearly and precisely in the context of the concrete signaling games we have been discussing. That the receiver evolves to do the right thing given the current state of nature requires that the sender's signal conveys information that faithfully communicates this state. When successful, the sender's signals, then, evolve to convey precisely the information needed for successful predictive action. Indeed, the *quantity* of information about the world communicated can be measured relative to a given partition of the states of nature.[20] The inquirers do not know precisely how their signals individuate the states of nature, but they know that they do so with sufficient precision to allow for the degree of successful prediction they have achieved. And when they eliminate old descriptive error or otherwise attain new sort of success in action, they know that their descriptive and predictive practice has in fact improved. They just do not know how.[21]

That said, descriptive and predictive *progress* over time is not guaranteed. It can happen that a new practice may not allow the inquirers to achieve a particular type of success they achieved in the context of their old practice, presumably because they got something else they wanted form the new practice. But when this happens, real inquirers are often willing to accommodate incommensurate, even incompatible, descriptions of the world at the same time. While quantum field theory answers to explanatory and predictive demands that Newtonian mechanics cannot in special situations, Newtonian mechanics is still used to accomplish the many descriptive, explanatory, and predictive tasks for which quantum field theory has proven itself entirely unsuitable. The thought is that we will be able to explain the predictive and explanatory successes of *both* theories at some point in the future. Even in the simple evolutionary models at hand, one can see how alternative characterizations of the nature might be radically incommensurable yet, from the god's-eye-view, entirely faithful given the descriptive tasks to which they answer. But, of course, in the context of real inquiry it is unclear how things would look from the god's-eye-view or even just from the perspective of our own future explanations of our current explanatory and predictive successes.

The evolutionary models we have been considering show how inquirers may fail to have the expressive resources to characterize how their descriptions get things right even as they in fact get things right. But even

[20] See Skyrms (2010) for a discussion of information transfer in signaling games. The partition that one uses to measure the information context of the signals might be stipulated from the god-eye-view. But the inquirer's might also measure the information content of their signals relative to the partition of states induced by their evolved descriptions. See the section on indifference in the other Barrett paper (2021) in this volume for more details regarding the properties of such induced partitions.

[21] Barrett (2008) discusses the sense in which one can know that one's descriptions of approximately true without know *how* they are approximately true.

when inquirers fail to know precisely *how* their evolved descriptive language provides faithful representations of their world, they can know that it *does* inasmuch as it allows for success action. The sort of successful action they manage is explained from the god's-eye-view by the inquirers having evolved faithful descriptions and discovering regularities. If the senders' descriptions did not reliably communicate information regarding the world or if the predictor had evolved the wrong predictive dispositions, their actions would not be uniformly successful.

7 An Endogenous Account of Epistemic Norms

The pragmatic notion of success and failure in action is central to the account of empirical inquiry proposed here. It is, hence, important to be clear about how such epistemic norms are to be understood.

Epistemic norms here are endogenous to the particular model. Specifically, the agents' epistemic norms are represented in their second-order dispositions to update their first-order dispositions to signal and to act. Hence *successful inquiry* is represented in the evolution toward a reflective equilibrium between the agents' first-order dispositions to signal and to act and their second-order dispositions *to update* their dispositions to signal and to act. Similarly, *descriptive error* is represented in the first-order dispositions to signal and to act that lead to actions that produce outcomes that trigger the agents' second-order dispositions *to change* their first-order dispositions. The model, then, relies on a clear endogenous notion of epistemic norms and how these norms track truth for the sake of successful action. On this account, the endogenous epistemic norms of human inquirers would, by analogy, be a function of the dynamics that determines our descriptive, explanatory, and predictive practice.[22]

The nature of *truth* in an evolutionary account of inquiry is a central theme in the companion paper in this volume, but there are two things

[22] In this regard, the present model shares much with the sort of belief-revision model of knowledge championed by C. S. Peirce. See, for example, Peirce's papers "The Fixation of Belief" and "How We Make our Ideas Clear" in Houser and Kloesel (1992).

that we can say at this point. First, while it is certainly the case that truth is a norm of inquiry, there is also an important sense in which it is also the other way around. We *take* a description of nature to be true when it is reflectively stable in the context of ongoing inquiry and action more generally. That is, a description that we take to be true is one that we believe will not in fact be revised when our descriptions are updated in light of our success and failure in action. It is something that we do not believe requires revision in order to avoid descriptive, predictive, and explanatory error.[23]

Second, while the inquirers' second-order dispositions, the reinforcing and punishing payoffs of the evolutionary game, determine what outcomes count as successful, they *do not determine* what actions in fact produce successful outcomes given the inquirers' endogenous norms. Rather, in order for the agents to be successful, their evolved dispositions to signal and to act must track *both* the second-order dispositions that determine what outcomes count as successful *and* the objective facts about the world that determine whether such an outcome will obtain. That is, while the agents' evolved signaling conventions and dispositions to act are entirely in the service of their second-order dispositions, *successful action* requires that their evolved descriptions of nature are in fact *faithful* and their evolved predictive dispositions are in fact *accurate* in the relevant sense.

8 Discussion

The present account illustrates how inquirers' predictive and descriptive dispositions may continue to coevolve over time yet nevertheless still faithfully describe the world. A predictive theory and its associated descriptive language are *together* in the service of the *de facto* values of the inquirers. The modeled inquirers cannot know precisely how their evolved descriptions get the world right. But they can know that they are getting something about the world right inasmuch as their shared success would

[23] See the companion paper Barrett (2021) for more details. See also Barrett (2008) for a general discussion of descriptive truth and the logical structure of inquiry.

be impossible without reliable representations of natural states and dispositions that in fact track regularities. In this sense, their success and failure in action provides them with an endogenous standard for true description — namely, they know that their descriptions correctly characterize the properties of the world they inhabit to a degree of precision that explains the degree of success they in fact enjoy in action. And they may even take a particular description to be true insofar as they consider it to be reflective stable under revision, but more on this in the companion paper.

From a god's-eye-view, one might concretely characterize precisely what it is that the inquirers are getting right in order to be successful. While the inquirer's themselves may never be able to do this at a time, given a future language with sufficient descriptive resources, they may be able to provide a backward-looking account of what they *were* getting right with the aim of explaining their past successes and failures in action by the lights of their *current* best description of the world. Given how they are situated in inquiry, such backward-looking endogenous accounts would be the closest the agents might get the sort of exogenous gods-eye-view account the modeler can give for their success.

Such agents can know that they have eliminated whatever descriptive error in fact led to prior failures in action. In this sense they know that they have made progress toward descriptive truth. To be successful they must get the world and how it behaves right in a sense that in fact leads to successful action given their aims. Their success might be described from a god's-eye-view by the correspondence truth of their descriptions, but the inquirers need not know how this works to know that their ongoing inquiry is allowing them to make progress toward the truth by the elimination of descriptive error.

The inquirers' endogenous account of their knowledge and the role it has played in their success and failure in action will invariably feel more pragmatic and less metaphysical then a god's-eye-view account. It would be a mistake for them to suppose, for example, that their success entails that the terms in their language refer to something like fundamental natural kinds that match up with canonical bits of the fundamental metaphysics of the world they inhabit. The right attitude for the inquirers to adopt will be that their language gloms onto *something* that is sufficient to explain the type and degree of success in action that they in fact enjoy and that they will be better placed to account for this success when they are able to avoid the descriptive errors that presently lead to unsuccessful action.

To possess a language that is in fact well-suited for describing the world for the sake of successful action is to know something significant about the world. The present account provides a first step in explaining how we might develop and learn to use such a language and how it might coevolve with our best theories and shifting predictive and explanatory demands.[24]

References

Alexander, J. M., Skyrms, B., & Zabell, S. L. (2012). Inventing new signals. *Dynamic Games and Applications, 2*, 129–145.

Argiento, R., Pemantle, R., Skyrms, B., & Volkov, S. (2009). Learning to signal: analysis of a micro-level reinforcement model. *Stochastic Processes and their Applications, 119*(2), 373–390.

Barrett, J. A. (2006). Numerical simulations of the Lewis signaling game: learning strategies, pooling equilibria, and the evolution of grammar. *UC Irvine Institute for Mathematical Behavioral Sciences Technical Report*. https://www.imbs.uci.edu/research/technical.php. Accessed 23 April 2020.

Barrett, J. A. (2007a). Dynamic partitioning and the conventionality of kinds. *Philosophy of Science, 74*, 527–546.

Barrett, J. A. (2007b). The evolution of coding in signaling games. *Theory and Decision, 67*(2), 223–237.

Barrett, J. A. (2008). Approximate truth and descriptive nesting. *Erkenntnis, 68*(2), 213–224.

Barrett, J. A. (2009). Faithful description and the incommensurability of evolved languages. *Philosophical Studies, 147*(1), 123–137.

Barrett, J. (2021). The evolution of truth and belief. In W. J. Gonzalez (Ed.), *Language and scientific research* (pp. 171–198). Cham: Palgrave Macmillan.

Barrett, J. A., & LaCroix, T. (2020). Epistemology and the structure of language. *Erkenntnis* (Forthcoming). https://doi.org/10.1007/s10670-020-00225-4.

[24] I would like to thank Brian Skyrms for many conversations on the topics of the paper and Travis LaCroix for helpful comments on an earlier draft.

Barrett, J. A., & Zollman, K. (2009). The role of forgetting in the evolution and learning of language. *Journal of Experimental and Theoretical Artificial Intelligence, 21*(4), 293–309.
Houser, N., & Kloesel, C. (eds.) (1992). *The Essential Peirce: Selected Philosophical Writings Volume 1 (1867–1893).* Bloomington and Indianapolis: Indiana University Press.
Kuhn, T. S. (1996). *The Structure of Scientific Revolutions.* Chicago: University of Chicago Press.
LaCroix, T. (2020). Communicative bottlenecks lead to maximal information transfer. *Journal of Experimental and Theoretical Artificial Intelligence* (Forthcoming). https://doi.org/10.1080/0952813X.2020.1716857.
Lewis, D. (1969). *Convention.* Cambridge, MA: Harvard University Press.
Quine, W. V. (1980). *From a Logical Point of View*, Cambridge, MA: Harvard University Press.
Skyrms, B. (2006). Signals. *Philosophy of Science, 75*(5), 489–500.
Skyrms, B. (2010). *Signals: Evolution, Learning, & Information*, New York: Oxford University Press.

5

Language, History and the Making of Accurate Observations

Anastasios Brenner

1 Introduction

Science could be defined as the art of making accurate observations. Indeed, observation taken in a broad sense — either as the description of phenomena occurring spontaneously in nature or as the registration of data produced by a technological setup — is an essential aspect of scientific activity. Yet the nature and significance of observation continues to raise debate. To be sure, post-positivists such as Thomas Kuhn, Imre Lakatos, Paul Feyerabend and Larry Laudan convincingly argued against logical positivists that there is no neutral language of observation, no clear-cut dichotomy between observation and theory, no absolute empirical basis. Some sort of consensus has been reached to the effect that facts are theory-laden. But the exact meaning of this assertion remains a controversial issue; all its consequences have not been drawn.

A. Brenner (✉)
Centre de Recherches Interdisciplinaires en Sciences Humaines et Sociales CRISES, Paul-Valéry Montpellier3 University, Montpellier, France
e-mail: anastasios.brenner@univ-montp3.fr

W. J. Gonzalez (ed.), *Language and Scientific Research*,
https://doi.org/10.1007/978-3-030-60537-7_5

To illustrate this point, let us take Lorraine Daston and Elizabeth Lunbeck's *Histories of Scientific Observation* (2011). They make the bold claim that "scientific observation lacks its own history" and proceed to initiate a research program aiming to fill in the gap by calling on new concepts and methods. This represents a major challenge to much scholarship deriving from post-positivists and has met with enthusiastic approval as well as strong criticism. It is part of a series of shifts, variously characterized as the practical turn, historical epistemology, integrated history and philosophy of science, etc.

Indeed, many researchers engaged in science studies today are calling for a change of direction, a move away from abstraction, formalization and generality toward the concrete, the pragmatic, the contextual. For a long time the overriding concern was with the formal structure of theory or at best a critical discussion of its elements. As regards observation, such an attitude led to a rather fragmentary, superficial, incomplete view. In contrast, I propose that we reexamine scientific observation, that we envisage it first and foremost as a practice, comprising different stages and operations: conception, execution, correction, approximation, repetition, etc. Observation appears then a complex process: one that evolves over time, one that is enriched by past experience.

Recourse has been made in philosophy of science either to logical analysis or historical study. Such an alternative echoes the divide between analytic philosophy and historical epistemology. There have been some attempts to combine the two traditions, but in the main philosophy of science has been dominated by formal methods. What is called for is a more determined effort to understand scientific activity. This means to reflect on the philosophical role that history is to play. One should be wary to consider it as a mere repository of discoveries or as an epistemological laboratory. The philosopher should draw freely on the techniques provided by the human sciences and the social sciences: history, sociology, anthropology, linguistics, etc.

I wish to make two preliminary remarks. First, in calling for a return to history, I shall inevitably be led to debunk some myths, to reveal limitations in past science. My intention is however not to question the achievements of science or to promote relativism. There may have been a tendency among students of history to exaggerate the contrast between the ancients and the moderns. I believe that to recall the complex historical process by which our knowledge of the world has been produced is rather to give substance, depth and credibility to the scientific enterprise. Experiment, to be fully understood, should be placed against the backdrop of the multiple attempts, the continuous effort, the directed attention, in sum, experience in the broad sense. This implies taking more seriously what scientists and philosophers of each epoch have to say about their activity, the events underway and the knowledge available at the time.

The second point is that I do not mean to reject altogether logical analysis. It can provide fruitful results in its own sphere, and an analysis of contemporary scientific language may well lead us to the study of past language. But logic cannot encompass history. Certain philosophically significant aspects of science can only be brought out by recalling the full record of the past.

The aim of this paper is to examine both the contributions and shortcomings of the new trends mentioned above with respect to our understanding of observation in the sciences. In particular, it appears important not to dissociate scientific practice from the epistemic values such as accuracy that infuse it.

2 Scientific Observation in Historical Perspective

I shall take as my point of departure the claim made by Lorraine Daston and Elizabeth Lunbeck in the collective volume they edited, *Histories of Scientific Observation*:

> Scientific observation lacks its own history: why? Countless studies in the history and philosophy of science treat one or another aspect of observation […]. But observation itself is rarely the focus of attention and almost never as an object of historical inquiry in its own right. Observation seems at once too ubiquitous, too basic, and altogether too obvious to merit a history. (Daston and Lunbeck 2011, 1)

This is a bold statement. The editors are claiming that earlier inquiry into the topic has missed the target. One could think of logical positivism, also called logical empiricism by its practitioners. The standard view of scientific theory that this movement proposed — as an axiomatic system comprising an observation language and a theoretical language linked by means of rules of correspondence — did not have much to say concerning the elaboration of theory or the practice of observation. The strict dichotomy between theory and observation, neutral observational terms, protocol sentences, simply beg the question. To be fair, these theses gave rise to controversies within the Vienna Circle, and Otto Neurath suggested a different perspective: "What is first given is our historical ordinary language with a multitude of imprecise, unanalyzed terms [*Ballungen*]" (Neurath 1983, 91). Rather than leading to a shift of doctrine, Neurath's arguments encouraged the search for more and more sophisticated logical solutions. More deeply, such an approach — positivistic — led to eliminate metaphysics, ontology and ethics. This would prove to be an impediment for the formulation of a satisfactory philosophical discourse on observation.

But Daston and Lunbeck are not satisfied either with the conceptions of post-positivists, such as Norwood Russell Hanson and Thomas Kuhn (Daston and Lunbeck 2011, 5). Although they both called on history, they did not really integrate historical study with philosophical reflection. To say that observation is theory-laden is to somewhat mitigate the dichotomy between theory and observation,[1] but it does not necessarily imply that one takes observation seriously.

Mainstream philosophy of science has been overly concerned with theory, when not simply formalization.[2] Little attention has been devoted to practice. Observation clearly involves the latter. Several difficulties of current philosophizing may find their source here. By taking into account observation, in the variety of its manifestations, one can broaden the scope to new fields, to new problems. The approach suggested by Daston and Lunbeck contributes both to what has been called the historical turn and the practical turn.

[1] For a view stressing the independence of theory and observation, see Kosso (1992).

[2] On the contrast between the analytic tradition and historical epistemology, see Brenner (2014).

What the editors provide us with is a study of observation from the Middle Ages up to the present. Our concept of scientific observation emerged slowly, and can be grasped only by a *longue durée* history. Originally, "observations" carried the meaning of "considerations" or "observances" and referred not to a particular area or type of activity but simply to common knowledge or experience. Only gradually did it acquire a specific focus, a definite direction. The authors contributing to *Histories of Scientific Observation* draw on a large number of fields stretching from the natural sciences to the human sciences. The studies zoom in on specific episodes taken from meteorology as well as physics. No hierarchy is prescribed; in particular, no attempt is made to reduce all scientific activity to the method of physics.

Let us focus on the study by Gianna Pomata, "Observation Rising: Birth of an Epistemic Genre, 1500–1650" (2011). Indeed, prior to the Scientific Revolution, one witnesses a shift: doctors, astrologers and astronomers moved further and further away from an Aristotelian approach to nature, one based on facts reported by authorities, most often rare or striking facts, that is experience or *empeiria*. These modern savants greatly increased repertories of data, they shared them and they encouraged looking for new results. Thereby, they were drawing on the skeptics of late Antiquity, in particular Sextus Empiricus. Not only did they put forth the term *observatio*, translating Sextus's *têrêsis*, but they borrowed two other related concepts: *autospia* and *phainomena*. Let us recall that *autopsy* means etymologically the examination one carries out oneself and that *phenomenon* was initially reserved for astronomy. Sextus favored this use for philosophical reasons: "phenomena are the result of an involuntary affect, whereas representations are the topic of controversy" (Sextus Empiricus 1933, I, 11 [22]). Thus Jean de Goris in his medical dictionary *Definitiones medicas*, published in 1564, states that "*têrêsis*" or "*observatio*" is a word of the empiric sect and he notes that the same goes for "*autopsia*." Pomata remarks furthermore that Henri Estienne, who gave an early translation of Sextus, follows this philosopher in using phenomenon in a broad sense. She summarizes then her results:

> The category of *observatio* first emerged in neo-Hippocratic medical circles with the recovery of the ancient empiric/Skeptic philosophical vocabulary, together with two other concepts, *autospia* and *phainomena*, also destined to play a fundamental role in the early modern renewal of the language of experience. The combined purport of these terms, in their ancient Empiric/Skeptic acceptation, was an emphasis on the distinction between direct experience (*autospia*) and indirect experience, the insistence on focused and repeated observation (*têrêsis*) as the foundation of empirical knowledge, and the urge to keep to the phenomena (*phainomenon*, or things as they appear), avoiding useless and contentious theorization. (Pomata 2011, 65)

One can gather from this careful historical study that observation is not something already given. It is the outcome of a slow and complex process, in which multiple cultural factors played a role. Moreover, the emphasis on observation in the modern sense came about in the peculiar context of the widespread reception of skeptical philosophy during the Renaissance.

It is helpful to complement this research on medieval and modern science by resorting to a comparison with ancient science. As Geoffrey Lloyd states, with respect to the development of empirical research in Antiquity: "The terms in which the ancients discussed some of the issues are, in important respects, different from those of the modern debate. In the classical period there is no exact equivalent, in Greek, for our observation" (Lloyd 1979, 129). Lloyd makes clear that there is a noteworthy difference in vocabulary between ancient science and modern science. This must be taken into account if we are to understand the nature of empirical research prior to the moderns.

Of course, Daston and Lunbeck's slender volume cannot claim to provide a definitive picture of its subject; the aim is rather to outline a program, to be completed with the help of other collaborators and future research.[3] One can point to numerous studies on the history of observational methods that pursue the path opened up. I shall endeavor here to fill in two obvious gaps. The criticism leveled at earlier conceptions is undoubtedly succinct. We need some notions of a history of philosophy of science. Furthermore, an observation becomes scientific on account of

[3] For two different reactions to this book, see Gingras (2012) and Salter (2012).

its accuracy. We likewise need to explore what is meant by this requirement. But I also believe that there is another dimension that has been left out, relating to the fact-value distinction, which has come again to the fore today: the ethical dimension of scientific observation.

3 Case Study One: Exact Observation and the Birth of Modern Science

Words are invented or redefined for the purpose of expressing theories and practices. Modern science was bound up with the making of a new language, not only a technical vocabulary, but a methodological, philosophical and cosmological discourse. Historians of science have gathered a wealth of material in this regard. But philosophers of science have been reluctant in the main to exploit this material, owing perhaps to Hans Reichenbach's rejection of the context of discovery, in accordance with his logical empiricism.

Three related terms come to mind in speaking of science: accuracy, precision and exactness. Let us pause to consider the meanings of these terms. "Precision" is a key concept in Charles S. Peirce's theory of science. He appropriated it to characterize a sign: "In those respects in which a sign is not vague, it is said to be definite, and also with a slightly different mode of application, to be *precise*, a meaning probably due to *praecisus* having been applied to *curt* denials and refusals" (Peirce 1998, 351). Peirce is calling attention to the etymology: *praecisus* comes from *caedere* with the prefix *prae*, signifying to cut off, to cut short; and it was employed in a technical sense in rhetoric. Obviously, our "precision" is far removed from the original meaning. The question then arises how the term evolved in such a way. A historical sketch is given by W. Hübener and S. Meier in Joachim Ritter's *Historisches Wörterbuch der Philosophie* under "Praecisio" (1989). The term is evidenced in a logical, semantic sense already in Thomas Aquinas, but what is of interest for our purposes is its use in the context of medieval natural philosophy. For example, in speaking of Venus, Mercury and the Sun, Nicole Oresme supposed several motions, of which some were specific for each celestial body:

> With this theory, it is not necessary to assume that the planets' being lower than or farther from the sovereign heaven causes their proper movement to be faster proportionally or in precise proportion [*selon proportionalité précise*], for in addition, we must consider and compensate for the force and will or desire of the motive power. (Oresme 1968, 128d)

Such formulations continued to be used during the fifteenth and sixteenth centuries. "Exactness," which lacks a corresponding substantive in classical Latin, derives from *exactus*, meaning to carry to completion. Our term is thus a late formation, which spread during the sixteenth and seventeenth centuries. Finally, "accuracy" is closest to classical Latin usage; "*accuratio*" designating carefulness. However, it came to acquire new meanings in the context of modern science.

Let us now examine in this light some significant instances. Copernicus's *On the Revolutions of Celestial Orbs* is considered to herald in modern astronomy. Recently, a critical edition has been published with complete biographical, historical and philological information. In the preface, Copernicus recalls the encouragement he received from the Catholic Church to publish his theory:

> For not so long ago, under Leo X, the Lateran Council considered the problem of reforming the ecclesiastical calendar [...]. From that time on, at the suggestion of that most distinguished man, Paul, bishop of Fossombrone, who was then in charge of this matter, I have directed my attention to a more accurate consideration [*accuratius observandis*] of these topics. (Copernicus [1992] 2015, 10; translation mine)

It is not obvious what force Copernicus is giving here to his "observations." Is he sacrificing to classical usage? Or does he mean something more for his novel theory? Going back from modern translations to the original, we note that the idea of precision is sometimes expressed with different words:

> For a long time, then, I reflected on this confusion in the astronomical traditions concerning the derivation of the motions of the universe's spheres. I began to be annoyed that the movements of the world machine, created for our sake by the best and most systematic Artisan of all, were not

> understood with greater certainty by the philosophers, who otherwise examined so precisely [*exquisite*] the most insignificant trifles of this world. ([1992] 2015, 7–8)

Literally, the passage reads as: earlier astronomers examined so "exquisitely," which is a strange turn of phrase for us today. The term is used later in the text in a technical sense to refer to the operation of correcting astronomical observations ([1992] 2015, 775). Copernicus also makes use of term "exact," in particular with regard to the phenomenon of the precession of the equinoxes ([1992] 2015, 183).

Despite the vagueness of Copernicus' language, a paradigm shift was underway: the move from the geocentric theory to the heliocentric theory. Copernicus was making a considerable effort to replace the traditional lexicon of medieval astronomy, which he had received from his schooling, with a new vocabulary. One finds in his book numerous borrowings from the Greek, terms with new meanings and neologisms. The modern discourse on science was in the making, but still inchoative.

This vocabulary was readily taken up by the followers of Copernicus. Such is the case of Galileo, and one can cite his *Sidereus Nuncius* of 1610 as the beginning of modern observational reports. The aim of the book is to provide a "history of the observations" that Galileo had recently carried out with the help of the telescope (Galilei 1992, 7). Galileo emphasizes repeatedly the exactness of his instrument, for example:

> In the first place it is necessary to obtain a most exact telescope [*Perspicillum… exactissimum*], one which represents the objects with their brilliance, distinctly, without any veil or mist, and which magnifies at least four hundred times. ([1610] 1992, 8; translation mine)

Only such a telescope, which we would call a precision instrument, made it possible to perceive what Galileo observed: the irregularities of the Moon, the satellites of Jupiter, star clusters, etc. He went on to measure the altitude of the lunar mountains as well as to estimate the periods of the orbits of the satellites of Jupiter. Thereby, he set an example for future research of a new kind, which strove to improve continually our methods of exploration.

Later, in his *Dialogue on the Two Chief World Systems* he recalls the observations of sunspots he had made with the help of the telescope in the following words:

> It came about that, continuing to make very careful [*diligentissime*] observations for many, many months, and noting with consummate accuracy [*accuratezza*] the paths of various spots at different times of the year, we found the results to accord exactly [*puntualmente*] with the predictions. (Galilei 1899–1909, VII, 379; trans. 1967, 352)

For him this is a strong argument in favor of the heliocentric hypothesis that he is defending: the sunspots are evidence of change in the superlunary world. In other passages Galileo calls on the term "exquisite," in a sense similar to that we noted above in Copernicus, where we would employ "precise" (1899–1909, VII, 256, 320, 415). We notice that Galileo was drawing on a much larger set of words than his modern translator. This is probably because these words did not have yet a sharp and well-established meaning. What Galileo was seeking to defend was his use of instruments and mathematics. Lurking behind such considerations are difficult philosophical problems concerning the relation of laws to experimental facts, the truth of mathematical representations and the mathematical structure of the world.

I have followed here Stillman Drake's English translation of Galileo's *Dialogue* for lack of better. First published in 1953, in the heyday of logical empiricism, this translation betrays a certain anachronism. Thus Floris Cohen, who published in 2010 a rich and substantial book, *How Modern Science Came into the World*, levels strong criticism at Drake's research on Galileo, emphasizing "some flaws that make its results unreliable:"

> Prominent among these flaws are Drake's at times urgent desire to present Galileo as, above all, a modern scientist making a complete break with all previous, inherently worthless philosophizing; a resulting, strangely flat picture with depth, tragedy, and even lifelikeness excised. (Cohen 2010, 218)

Cohen proposes to follow an altogether different line of interpretation based on historical erudition and careful textual reading, as represented by Alexandre Koyré, E. J. Dijksterhuis and R. S. Westfall. He cites with approval Westfall's synopsis concerning Galileo's conception of nature: "That it was an impossible amalgam of incompatible elements, born of the mutually contradictory world views between which he stood poised" (Cohen 2010, 218).

The term accuracy in the sense of "precision and correctness" enters the English language towards the middle of the seventeenth century. One of the earlier occurrences can be found significantly in Newton's Preface to the first edition of the *Mathematical Principles of Natural Philosophy*. "Accuracy" is employed here to characterize the distinctive trait of the science that is being expounded. To provide a complete commentary, one would have to spell out the rhetorical techniques, polemics and allusions that the author deploys. While calling on the authority of Pappus, Newton seeks to break with tradition. His discourse can be connected with the Quarrel of the ancients and the moderns, the Battle of the books. But what is of interest for us here is the term accuracy, which occurs several times at the beginning of the Preface. Newton proceeds to reverse the precedence of geometry over mechanics: "Geometry is founded on mechanical practice, and it is nothing other than that part of universal mechanics, which reduces the art of measuring to exact [*accurate*] propositions and demonstrations" (Newton 1687, XV; translation mine). Newton is of course promoting his theory. This does not mean that earlier science did not seek rigor and certainty, but undeniably new standards were being set, and these standards had an impact on our understanding of precision. Accuracy, as illustrated by Newtonian method, came to characterize science.

After Newton the realm of accuracy expanded constantly from mechanics proper to numerous physical phenomena such as heat, electricity and magnetism. New areas came to be submitted to quantitative methods: chemistry, biology and sociology. In consequence, the question of distinguishing exact science from other forms of scientific knowledge sprang up, and the expression "exact sciences came into being. Finally, in the nineteenth century, as the Newtonian paradigm entered a state of crisis, a more reflexive, philosophical analysis of the nature of scientific practice arose.[4]

4 Case Study Two: Observation in the Life Sciences

Let us now turn to the case of scientific observation in the life sciences. It is interesting in this respect to examine a few passages from Claude Bernard's *Introduction to the Study of Experimental Medicine*, published in 1867. He presents his reflections on science as a scientific philosophy, in an effort to distance himself from Auguste Comte's positivism. Bernard opens his book with the characterization of observation as distinguished from experiment: "The word observation in the singular, in its general and abstract use, means noting a fact accurately [*la constatation exacte d'un fait*] with the help of appropriate studies and means of investigation" (Bernard 1966, 40; trans. 1957, 11). He emphasizes the specificity of medicine:

> Medical investigation is the most complicated of all [studies]: it includes all the methods proper to anatomical, physiological and therapeutic research, and, as it develops, it also borrows from chemistry and physics many means of research which become powerful allies. (Bernard 1966, 43; trans. 1957, 14)

Bernard is intent on making a case for the experimental method in the life sciences.

He goes on to explicate the conditions of this method:

> The first condition to be fulfilled by men of science, applying themselves to the investigation of natural phenomena, is to maintain absolute freedom of mind, based on philosophic doubt. Yet we must not be in the least sceptical; we must believe in science, i.e., in determinism; we must believe in a complete and necessary relation between things, among the phenomena

[4] For an overview of rational criteria or epistemic values, see Brenner (2011).

> proper to living beings as well as in all others; but at the same time we must be thoroughly convinced that we know this relation only in a more or less approximate way, and that the theories we hold are far from embodying changeless truths. (Bernard 1966, 68–69; trans. 1957, 35)

Bernard thus summarizes his conception of experimental method. One gathers that this practice implies several conditions, which are external or transcendental to science: freedom of thought, philosophical doubt and belief. One should read the whole paragraph from which this quote is taken, which carries the title: "Experimenters must doubt, avoid fixed ideas, and always keep their freedom of mind" (1966, 68; 1957, 35). Underlying Bernard's account is the contrast between science and religion.

Bernard is defending here the superiority of science with respect to religious or philosophical discourse. For science alone enables us to rid ourselves of the prejudices of childhood as well as to keep free from discourses which bear only on words and not on reality. Let us notice that science implies certain moral qualities:

> If men discuss and experiment [...] to prove a preconceived idea in spite of everything, they no longer have freedom of mind, and they no longer search for truth. Theirs is a narrow science, mingled with personal vanity or the diverse passions of man. Pride, however, should have nothing to do with all these vain disputes. (Bernard 1966, 72; trans. 1957, 39)

Reference is made here to the two meanings of the concept of vanity, an eminently religious concept: pride and futility. This passage on the ethics of experimental practice draws explicitly on Francis Bacon and his *Novum Organon*. But the epistemic values mentioned there must be adapted to the science of the nineteenth century and the particular case of medicine. The sphere of knowledge had expanded from physics to chemistry and finally to biology, proceeding up through the classification of the sciences. Science had displaced religion first with respect to the material world and then with respect to the human subject. Bernard belongs to the effort to reformulate the conceptions of science during the nineteenth century. He took up and remodeled the Kantian distinction between objectivity and subjectivity. He was not alone, Hermann von Helmholtz in Germany and Thomas Henry Huxley in England were proposing similar conceptions.

Scientific observation is the result of a long and demanding training, which implies abandoning one's opinions and setting aside one's passions. The scientist becomes a machine for registering data. Or in Bernard's terms:

> Observers, then, must be photographers of phenomena; their observations must accurately [*exactement*] represent nature. We must observe without any preconceived idea; the observer's mind must be passive, that is, must hold its peace; it listens to nature and writes at nature's dictation. (Bernard 1966, 52; trans. 1957, 22)

Claude Bernard's research involved experimenting on animals, in particular dogs, and the aim was to understand the functioning of the human body, so as to be able to make progress in medicine. This inevitably led to some concerns, the need to examine the ethical implications of scientific research, a point to which I shall return later.

5 Philosophy of Science and the Role of Ethics

Let us now attempt to draw the consequences of these two case studies. As we saw, after Newton, the requirement of exactness had come to characterize science. Such was to be the driving aim of research. The question then arose of the extension of this requirement. Newtonian science spread through the entire community of savants during the eighteenth century. By the time d'Alembert wrote his "Preliminary Discourse" to the *Encyclopedia*, exactness had come to designate science in its most exemplary form. He was among those who promoted the expression "exact sciences," employing it in particular to admonish those like the poet Boileau, who defended the ancients against the moderns in the famous quarrel. (D'Alembert [1779] 1821, 414).

But several questions remain to be answered: how to understand the application of mathematics to the external world or, in other words, the relation of an absolutely exact and infinitely precise language to the concrete, complex, fluctuating properties of the universe. Such was the object of philosophy of science, which became an autonomous field of inquiry during the nineteenth century. Measurement implied precision instruments, their construction and use. It became routine to vary the methods of measurement and to combine them in view of greater precision. Each instrument required its own theory, explaining its functioning, yielding an estimate of its degree of precision, making possible the correction of its values.

Human perception also came under rigorous study. Its particularities and limits were brought to light. Let us take the case of the philosopher-scientist Ernst Mach; he pointed out that our ear is but an instrument, which can be submitted to the basic laws of science, and he conceived a psychophysics that would bridge the gap between the two disciplines. Mach produced numerous historical studies of various concepts and branches of physics. His most famous work, *The Science of Mechanics*, bearing the subtitle *A Critical and Historical Account of Its Development* (Mach [1883] 1974), provided an abundant documentation on the history of experimental techniques, and Mach did not fail to mention as well some apparatuses that he had devised for his own teaching.[5]

He reflected on the procedures of classical science, emphasizing the metaphysical presuppositions that infused early modern science. First of all, it should be noted that "exactness" carries connotations of "truth" or "reality." By an exact or accurate theory, we might be tempted to infer an agreement with the facts, that is one for which the predictions match the observations. But this is to assume a correspondence theory of truth: an adequation between the mental representation and the physical reality.

How did Galileo bridge the gap between the mathematical laws he formulated and his experimental results? Concerning his law for free fall, he was sticking to the methods of ancient geometry; he had at his disposal neither analytic geometry nor integral calculus. Likewise, no clocks measuring seconds were available. He counted his pulse beats or collected the water flowing from a bucket with a hole, while a ball descended on an inclined plan. The imprecision is of course glaring by our standards. In

[5] For more on Mach's Historical Approach, see Brenner (2019).

order to counter objections, Galileo resorted to a general principle: "Nature is accustomed to act by employing means of the most ordinary, most simple and easiest sort" (Galileo [1638] 1899–1909, VIII, 197). He further called on God qua mathematician.

Mach was quick to condemn such phraseology. The scientist of the nineteenth century, claiming to be objective, should clearly distance himself or herself from such extraneous speculations. As he writes:

> Men generally are inclined to hypostatize their abstract ideas, and to ascribe to them a reality outside consciousness. Plato, in his doctrine of Ideas, only made a somewhat free use of this tendency. Even inquirers of the rank of Newton, despite their principles, were not always careful enough in this respect. (Mach [1886] 1986, 56)

It is important to understand how exactness is in fact obtained in scientific practice. Our sensations concerning physical properties are replaced by symbols, which only partially correspond to the former. These make possible measurement and mathematical manipulation. But one should be careful not to infer from the initial sensation an actual or real property of which our symbol would be a more or less inexact expression.

Mach goes on to refer us to his critical analysis of Newton's conception of absolute space and time. There is no empirical foundation for such notions, and the physicist should now dispense with them entirely. We have no physical means to detect the properties of the medium in which bodies are supposed to have absolute motions. Newton's recourse to the *sensorium* of God is a subterfuge. It sparked off a controversy with Leibniz over the appropriate or righteous conception of divinity. Such metaphysical considerations are surely beyond the bounds of science.

A second point should be made with regard to our case studies. It was medical doctors who awoke philosophers from their dogmatic slumber. The development of new therapeutic techniques made a breakthrough by the mid-twentieth century. It raised new hopes for vanquishing disease, but at the same time provoked fears. In response a whole field of applied ethics was brought into existence: bioethics, environmental ethics, professional ethics, etc. A long battle had to be waged. For ethics had been for the most part eliminated from philosophy *per se*. This was one of the unfortunate consequences of logical positivism and early analytic philosophy. Otto Neurath considers "ethics and jurisprudence as metaphysical residues" (Neurath 1983, 78). Alfred Jules Ayer, who embraced with enthusiasm the program of the Vienna Circle, notoriously advised philosophers to give up all "moralizing" and to devote themselves to metaethics, in other words to the meaning of moral terms, the logical relation between moral judgments and other forms of judgments as well as the epistemic status of moral judgments. Such a move is basically to dispense with any ethical substance. It took a long time to reinstate a genuine ethical discourse in philosophy. The reappearance of such a discourse only took place in the late sixties, at a time when logical positivism was loosing ground under the criticism of the historical school.

6 Conclusion

Daston and Lunbeck, among other thinkers belonging to the movement of historical epistemology, have opened up a whole new field of exploration. It remains to draw from the rich and detailed studies of past science on offer the philosophical consequences involved. Indeed, the relation between observation and theory allows for no sharp dichotomy. Observations become possible within a certain context, determined by instruments, concepts and goals. It is then necessary to inquire into the historical conditions of possibility of scientific activity, to explore the archives of science.

Such an approach encourages us to call on a broad spectrum, namely to adopt a multiplicity of methods, to resort to different disciplines, to take into account what the practicing scientist has to say explicitly or implicitly about his or her activity. Exact observation involves more than meets the eye. Exactness represents a whole array of methods, practices and attitudes. As a general category applying to all sciences, to their numerous methods of exploration, it is left undefined. In this sense one could say paradoxically that exactness as a concept is vague.

It is time now to answer an objection that could be raised. The study carried out above has taken the concept of exactness and broken it down into a variety of attitudes, notions, operations — a procedure that may seem to undermine what is a constitutive aspect of science. This would be to forget what historical investigation has to offer: the documents, the testimonies, the specimens. To study the history of a discovery is to bring to memory the substance of scientific practice. Moreover, one may call on a series of disciplines relevant to the issue of understanding the human endeavor of science, that is the social sciences. These are indeed sciences in their own right. By bringing them to bear on our topic, we have simply drawn on one group of sciences to explain another; in a sense we have set into motion the encyclopedia of knowledge. Such is hardly a relativistic option.

References

Bernard, C. (1966). *Introduction à l'étude de la médecine expérimentale* (1865). Paris: Garnier-Flammarion. English translation by Henry Copley Greene (1957). *An introduction to the study of experimental medicine*. New York: Dover.

Brenner, A. (2011). *Raison scientifique et valeurs humaines: Essai sur les critères du choix objectif*. Paris: Presses Universitaires de France.

Brenner, A. (2014). Epistemology historicized: The French tradition. In M. C. Galavotti, D. Dieks, W. J. Gonzalez, S. Hartmann, T. Uebel, & M. Weber (Eds.), *New directions in the philosophy of science* (pp. 727–736). Dordrecht: Springer.

Brenner, A. (2019). Mach, Duhem and the historical method in the philosophy of science. In F. Staedler (Ed.), *Ernst Mach: Life, work, influence* (pp. 637–650). Dordrecht: Springer.

Cohen, F. (2010). *How modern science came into the world*. Amsterdam: Amsterdam University Press.

Copernicus N. ([1992] 2015). *De revolutionibus orbium cœlestium*. Paris: Les Belles Lettres (Latin original and French translation). English translation by Edward Rosen, Baltimore: Johns Hopkins Press.

D'Alembert, J. Le Rond ([1779] 1821). *Éloges*. In *Œuvres* (Vol. II, pp. 351–440). Paris: Belin.

Daston, L., & Lunbeck, E. (2011). *Histories of scientific observation*. Chicago: University of Chicago Press.

Galilei, G. ([1610] 1992). *Sidereus nuncius*, edited by I. Pantin. Paris: Les Belles Lettres.

Galilei, G. ([1632] 1899–1909). *Dialogo sopra i due massimi sistemi del mondo*. In A. Favaro (Ed.), *Le Opere* (Vol. VII). Florence: G. Barbera. English translation by Drake Stillman (1967), *Dialogue concerning the two chief world systems*. Berkeley: University of California Press.

Galilei, G. ([1638] 1899–1909). *Discorsi e dimostrazioni matematiche intorno a due nuove scienze*. In A. Favaro (Ed.), *Le Opere* (Vol. VIII). English translation by Drake Stillman (1974), *Two new sciences*. Madison, WI: University of Wisconsin Press.

Gingras, Y. (2012). Review of *histories of scientific observation* by Lorraine Daston & Elizabeth Lunbeck. *Isis, 103*(1), 157–158.

Hübener, W., & Meier-Oeser, S. (1989). Praecisio. In J. Ritter et al. (Eds.), *Historisches wörterbuch der philosophie* (Vol. VII, pp. 1211–1218). Basel: Schwabe.

Kosso, P. (1992). *Reading the book of nature*. Cambridge: Cambridge University Press.

Lloyd, G. (1979). *Magic, reason and experience*. Cambridge: Cambridge University Press.

Mach, E. ([1883] 1974). *The science of mechanics* (T. J. McCormack, Trans.). Chicago: Open Court.

Mach, E. ([1886] 1986). *Die Principien der Wärmelehre*. English translation by T. J. McCormack, *Principles of the theory of heat*. Dordrecht: Reidel.

Neurath, O. (1983). *Philosophical Papers 1913–1946* (R. Cohen & M. Neurath, eds.). Dordrecht: Reidel.

Newton, I. ([1687] 1972). *Philosophiae naturalis principia mathematica* (I. A. Koyré & I. B. Cohen, eds.). Cambridge, MA: Harvard University Press. English translation by I. B. Cohen & A. Whitman (1999). *The principia: Mathematical principles of natural philosophy*. Berkeley: University of California Press.

Oresme, N. ([1377] 1968). *Le livre du ciel et du monde* (A. D. Menut, ed.). Madison: University of Wisconsin Press.

Peirce, C. S. (1998). Issues of pragmaticism. In N. Houser et al. (Eds.), *The essential Peirce* (Vol. II). Bloomington: Indiana University Press.

Pomata, G. (2011). Observation rising: Birth of an epistemic genre 1500–1650. In L. Daston & E. Lunbeck (Eds.), *Histories of scientific observation* (pp. 39–80). Chicago, IL: University of Chicago Press.

Salter, A. (2012). Review of *Histories of scientific observation* by Lorraine Daston and Elizabeth Lunbeck. *HOPOS: The Journal of the International Society for the History of Philosophy of Science, 2*(1), 196–200.
Sextus Empiricus. (1933). *Outlines of pyrrhonism* (R. G. Bury, Trans.). Cambridge, MA: Harvard University Press.

Part III

Scientific Language in the Context of Truth and Fiction

6

The Evolution of Truth and Belief

Jeffrey Barrett

1 Introduction

Our descriptive language coevolves with our beliefs and commitments. We use our best evolved language to describe the world. But we also use it to characterize our beliefs and commitments. Among other things, the ability to describe one's beliefs allows one to communicate one's conditional dispositions to others. And the ability to indicate one's commitment to the truth of a description or to one's degree of belief in it allows one to indicate something regarding the proper use of that description for the purpose of successful action.

We are concerned here with how one's beliefs and commitments might coevolve with the descriptive language one uses to characterize them. As in the companion paper in this volume (Barrett 2021), we will model this evolutionary process using signaling games, but here we will consider a yet more general variety of signaling game. While in the tradition of Lewis (1969) signaling games, *generalized signaling games* are significantly more subtle. Most saliently for our purposes here, such games may have

J. Barrett (✉)
Department of Logic and Philosophy of Science, School of Social Sciences, University of California at Irvine, Irvine, CA, USA
e-mail: j.barrett@uci.edu

W. J. Gonzalez (ed.), *Language and Scientific Research*,
https://doi.org/10.1007/978-3-030-60537-7_6

a hierarchical structure. We will consider how they work by discussing a number of concrete examples.[1]

As with a simple signaling game, a generalized signaling game might self-assemble from the ritualization of individual actions of the agents and the behavior of the agents is tuned in the context of an ongoing evolutionary process.[2] The evolutionary process in the present models will be characterized by a learning dynamics. We will refer to the parts of these models as agents or inquirers, but they may also be understood as functional components of a larger social group or of an individual agent. And, just as in the companion paper, the learning models might here be translated into population models with an evolutionary dynamics describing how the distribution of types in the population change over time.

We will begin by considering how a notion of *truth* might coevolve with an associated descriptive language. The particular notion of truth that evolves will depend on the details of the game. When successful, the notion of truth one ends up with will have evolved to play a role in the game. One might individuate alternative notions of truth by the games in which they might coevolve and the roles that they come to play in those games.

We will then consider how agents might evolve a language that allows them to characterize their beliefs and degrees of belief. More specifically, we will consider how agents might evolve a language that they might then use to characterize their own dispositions to act. Here an agent learns to represent the predictive dispositions of other agents then learns to apply the evolved representational capacities to herself.

Finally, we will consider evolutionary conditions under which a principle of indifference that assigns equal prior probabilities to each possible state of the world as partitioned by one's evolved descriptive language would be successful. This does not justify the use of a principle of indifference generally. Rather, it explains how an evolutionary process might render such a principle reliable under special circumstances.

---

[1] See Barrett and Skyrms (2017) and Barrett, Skyrms and Cochran (2020) for examples of hierarchical games and the former paper for a discussion of how such games might form.

[2] See Barrett and Skyrms (2017), Barrett, Skyrms and Mohseni (2018), and Barrett (forthcoming) for detailed discussions of how evolutionary games might self-assemble.

Each of the stories we consider here is a small chapter in the larger epistemological story of saying how agents might coevolve a useful descriptive language, knowledge about the world they inhabit expressed in that language, and the ability to use the language for reflective deliberation. In short, these stories are in service of providing a pragmatic account of language and inquiry.

2 Pragmatic Notions of Truth

To say that a description is true is to say that there is something right about it, that it is in some sense a *proper* description. Thomas Kuhn suggested one standard of proper description in his description of the choice between competing scientific paradigms in the context of inquiry:

> Like the choice between competing political institutions, that between competing paradigms proves to be a choice between incompatible modes of community life. …When paradigms enter, as they must, into a debate about paradigm choice, their role is necessarily circular. …[W]hatever its force, the status of the circular argument is only that of persuasion. …As in political revolutions, so in paradigm choice—there is no standard higher than the assent of the relevant community. (Kuhn 1996, 94)

On this view, the highest standard for accepting a paradigm's description of the world is the *assent of the community of inquirers*. A description of some aspect of the world is *proper*, then, when it reflects the shared consensus of the community. Insofar as one associates truth with proper description here, one might take a description to be *true* when it represents the shared consensus of the community of inquirers. One might think of this approach as providing a weak coherence notion of truth, one where the sort of coherence at stake is socially determined consensus in the context of inquiry.

The American pragmatist C. S. Pierce was committed to a somewhat stronger notion of proper description. For Peirce, "The opinion which is fated to be ultimately agreed to by all who investigate, is what we mean by the *truth*, and the object represented in this opinion is the *real*" (Peirce

1878, p. 139 in "How to Make Our Ideas Clear" in Houser and Kloesel (1992). While the first clause may look something like Kuhn's notion, Peirce's account of proper description is importantly different in how it ties the *truth* to the *real* by means of *successful action*. For Peirce, the community of inquirers will reach the truth because they will work to eliminate descriptive error for the purpose of successful action. The *real* is what explains success and failure in action. It explains why consensus can never be taken to be constitutive of the truth and in doing so explains how the entire community of inquirers might be wrong and, consequently, unsuccessful in action, even by their own lights. This is something that would be impossible if Kuhn were right that here is no standard higher than the assent of the relevant community.

One might take the proper use of a descriptive expression to be given by the evolved standards of the community or by what has been found through empirical inquiry to lead to successful action or by something else. In the two examples above, the alternative notions of proper description might be thought of as individuating alternative pragmatic notions of truth. The more general thought is that one might distinguish between alternative pragmatic notions of truth by distinguishing between alternative notions of proper description, which one might model by the roles that one's descriptions come to play in the context of particular evolutionary games. In this way, evolutionary games provide a way of both individuating pragmatic notions of truth and explaining how such notions might evolve.

Before one has a descriptive language, there are no descriptions that might be true or false. As one's descriptions evolve meanings, one might coevolve the ability to characterize some descriptions as true and others as false. We will consider how alternative notions of truth might coevolve with a descriptive language and how such notions might be distinguished by what they have evolved *to do*, by the roles they have come to play in the games in which they have evolved. Inasmuch as one takes the *meaning* of a description to be given by the circumstances under which one would judge it to be *true*, alternative pragmatic notions of truth also provide alternative pragmatic notions of descriptive meaning.

We will consider two composite signaling games where different pragmatic notions of truth might evolve. The first represents a notion of *reflective truth* that tracks conventional use. This might be thought of as a weak coherence notion of truth similar to the sort suggested by Kuhn's standard of proper use. The second represents a notion of *reliable truth* that tracks use for the purpose of successful action. This is a stronger notion of truth of the sort suggested by Peirce. In each case we will see how a relatively weak notion of descriptive truth might track a stronger notion of descriptive truth. In particular, we will consider how a weak coherence notion of truth might track a strong coherence notion of truth which might, in turn, track a yet stronger correspondence notion of truth. Along the way, the aim is to consider how one might characterize alternative notions of truth by the role they play in ongoing inquiry.

Each of the composite games we consider here consists of two Lewis-Skyrms signaling games interacting in a hierarchical way.[3] One of these is a *base game* that evolves a language descriptive of nature. The other is a *metagame* that coevolves a language descriptive of the base game and its evolving language.

2.1 Basic Signaling, Truth, and Inquiry

Let's start by considering the first signaling game we discussed in the companion paper but with a different learning dynamics. The basic structure of the game is represented in Fig. 6.1. The sender observes the state of nature, then sends a signal. The receiver, who cannot observe nature, then performs a predictive action (like bring a hat to the picnic) that is successful if and only if it matches the state of nature (bringing the hat is successful when it is sunny). We will assume here that both agents learn by *simple reinforcement*. Here if the receiver's act is successful, then the disposition that led to each agent's action on the current play of the game is reinforced. On simple reinforcement learning, a ball is added to each of the urns from which the agent drew of the type that they drew to represent what they learned from the successful play. This increases the chance that they will draw a ball of that type in the future under similar circumstances. If the agents are unsuccessful, they just return the balls they drew

[3] See Lewis (1969) and especially Skyrms (2010) for examples of the sort of basic games from which hierarchical games are constructed.

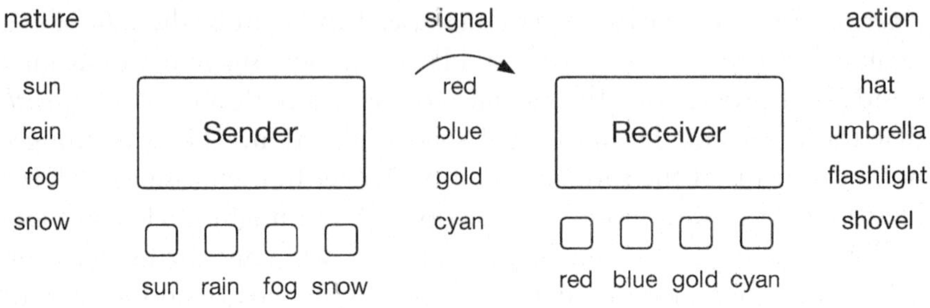

Fig. 6.1 A basic signaling game (Source: Author)

to the urns from which they drew them. Note that there is no upper bound with this type of learning on the number of balls of a given type in each urn and the agents never discard balls from the urns.

What happens on this learning dynamics is a little different from what we saw in the companion paper (Barrett 2021). On simulation the agents start by randomly signaling and acting. But here, as they learn from their successes, they evolve a nearly perfect signaling system, where each signal produces a successful predictive action given the state of nature, only about three-quarters of the time. Otherwise, they evolve a suboptimal partial-pooling equilibria involving mixed strategies that are successful (as it happens) just under three-quarters of the time. These suboptimal cases will play an important role in the story later.

Such simple signaling games may come to interact with each other and self-assemble more complex games. One way that this might happen is by modular composition. Since we are interested here in how a truth predicate might coevolve for a simple signaling language like the one just described, we will consider a composite hierarchical game where a *base game* takes states of nature as input and a *metagame* takes states of nature and states of the base game as input. If both parts of the composite game are successful, then (1) the base game will evolve a language that is descriptive of nature and (2) the metagame will coevolve a language that is descriptive of the base game and its descriptive language.

Inquirers who evolve successful interrelated descriptive and predictive dispositions in a signaling game like the one we are considering here do not know precisely how their evolved descriptions connect to the world.

But their predictive success provides them with evidence that they are getting something about the world right. That they are getting something right is what explains their success from the god's-eye-view.[4]

We are interested here in how such inquirers might coevolve an endogenous notion of truth that works with their base language to capture the sense in which their base language descriptions are getting the world right. There are a number of different evolutionary games one might consider. Different evolutionary games typically individuate different pragmatic notions of truth.

In the first game we consider, the notion of truth that evolves tracks the establishment of conventions in the base language as the agents' dispositions evolve toward a reflective equilibrium. The second tracks whether descriptions in the base language in fact lead to reliable action. In both cases the associated notions of truth, when they evolve, come to track endogenous norms. They sometimes also evolve to track descriptive truth from the god's-eye-view.[5]

2.2 Reflective Truth

Reflective truth is a variety of conventional truth. It represents a particular type of success in inquiry. The present game involves learning to copy the actions of other language users when those actions reflect the established practices of the community. When each part of this composite game is successful, reflective truth will also track correspondence truth. More specifically, when successful, the agents will evolve a metalanguage that characterizes when the base language is being used in a way that is in fact descriptive of the world given the evolved meanings of the base language expressions.

As discussed in the companion paper, success in inquiry may be understood as establishing a reflective stability between the agents' basic and

[4] This point is discussed in detail in the companion paper (Barrett 2021) in this volume. See also Barrett (2008) for a discussion of the nature of judgments regarding truth and approximate truth on a belief-revision model of knowledge.

[5] See Barrett (2016, 2017) for complementary discussions of the evolution of truth and probability predicates.

higher-order dispositions. One might take the agents' epistemic norms in the context of a signaling game to be represented in their second-order dispositions to update their first-order dispositions to signal and to act. *Successful inquiry* then is when the composite system evolving toward a reflective equilibrium between the agents' dispositions to signal and to act and their dispositions to reinforce their dispositions to signal and to act. If the agents do evolve stable descriptive and predictive dispositions, then they are doing the best they can to satisfy their second-order dispositions given where they started and that they are learning using precisely those second-order dispositions.[6] Inquirers may evolve a metalanguage for representing this endogenous norm of reflective stability if they can come to successfully track and describe whether they are in fact evolving stable descriptive dispositions.

A pragmatic notion of reflective truth that tracks conventional usage might coevolve with a basic descriptive language. One way that this might happen is if the metagame sender learns to track and to communicate whether the base-game sender just used whatever has to this point evolved to be the conventional signal for the current state of nature in the most recent play of the base game. We will consider how such a metagame might coevolve a notion that represents the *customary use* of the base-game's evolved signals. We will also consider how this endogenous notion may track stronger varieties of truth as well.

The structure of the reflective truth game is given in Fig. 6.2. Here the base game works precisely as described above, but the play of this game is observed by the metagame sender. The metagame sender observes whether the base-game sender sends the signal in the current play that she has used most often given the current state of nature. One metagame receiver action is successful if the base-game signal was in fact the most often used given the current state of nature. If the base-game signal was not the most often used given the current state of nature, then the other metagame receiver action is successful. We will suppose that the metagame agents, like the base game agents, learn by simple reinforcement. To be

[6] As we saw in the companion paper (Barrett 2021), that the agents reach such a reflective equilibrium does not mean that they have evolved optimally reliable practices. We will return to consider the issue of reliability more directly in the next model.

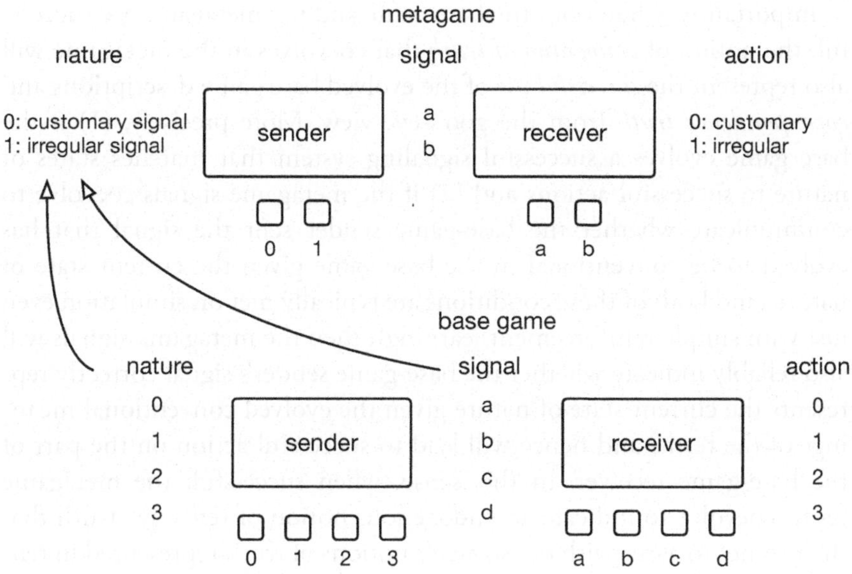

Fig. 6.2 Reflective truth (Source: Author)

successful, the metagame agents must learn to describe whether the base game is being played in the customary way given its evolution so far.

On simulation both the base game and metagame agents begin by randomly signaling and acting. But as the base game agents evolve a basic signaling language, they establish conventions that the metagame agents learn to track and represent. Specifically, the metagame agents typically (0.78) coevolve signals that distinguish between the base-game sender sending the signal that she has sent most of the time in the current state and not.[7]

The upshot is that when both games are successful, the base game agents evolve a simple language that they use for reliable description and the metagame agents evolve a simple language that the metagame agents use to describe how the base game language is being used. Specifically, the metagame language evolves to express whether the base game language is currently being used in what has evolved to be the conventional way.

[7] This is again with 1000 runs and 1×10^6 plays per run.

Importantly, when both the base game and the metagame are successful, the notion of *conventional truth* that coevolves in the metagame will also represent the *successful use* of the evolved base game descriptions and *correspondence truth* from the god's-eye-view. More precisely, (1) if the base game evolves a successful signaling system that matches states of nature to successful actions and (2) if the metagame signals coevolve to communicate whether the base-game sender sent the signal that has evolved to be conventional in the base game given the current state of nature (and both of these conditions are typically met on simulation even just with simple reinforcement learning), then the metagame signals will also reliably indicate whether the base-game sender's signal correctly represents the current state of nature given the evolved conventional meanings of the terms and hence will lead to successful action on the part of the base-game receiver. In this sense, when successful, the metagame terms coevolve to indicate an endogenous notion of reflective truth that then comes to agree with the stronger notions of truth represented in reliable use and the correspondence of the base-game expressions to matters of fact from the god's-eye-view.

2.3 Reliable Truth

An endogenous notion of *reliable truth* might coevolve more directly with a basic descriptive language in the context of different sort of evolutionary game on simple reinforcement learning. One way that this might happen is if the metagame sender directly monitors the conditional success rates of actions resulting from the use of the evolving base-game expressions. In the present model the metagame agents evolve a language that allows the metagame receiver to copy the actions of the base game receiver but only when the current basegame signal leads to reliable action.

The structure of the reflective truth game is given in Fig. 6.3. The base game here is a two-state signaling game with simple reinforcement learning. As in the earlier basic signaling games, the sender sees a random, unbiased natural state then sends signal *a* or *b*. Then both base-game agents are rewarded if the receiver's action matches the state of nature. But this time we will suppose that the base game is noisy, and the receiver

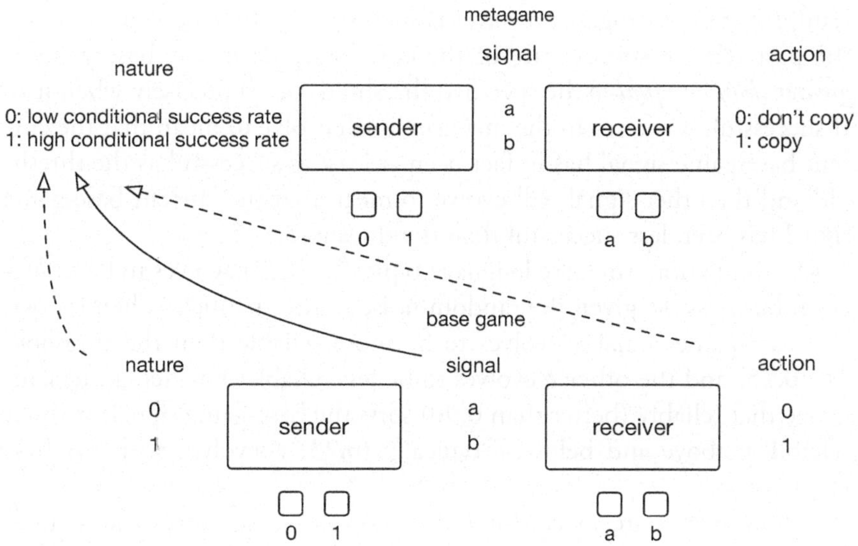

Fig. 6.3 Reliable truth (Source: Author)

hence does not always correctly hear the sender's signal. More specifically, we will suppose that with probability 0.5 the sender's signal is simply set to *b* on each play of the game.

The metagame here is also a two-state signaling game with simple reinforcement learning. The sender observes whether the current base-game signal has a cumulative success rate greater than or equal to a threshold *T* (which we take to be 0.75 for the purpose of having something concrete to start with). One metagame receiver action (the signal is *probably reliable*) is successful if the cumulative success rate of the current signal is in fact greater than or equal to *T*; otherwise, the other action (the signal is *probably not reliable*) is successful. Success of the metagame, then, depends on learning to distinguish between base-game signals that have been reliable and those that have not and coevolving a descriptive metalanguage that communicates this information.

The metagame sender observes the evolving conditional success rate for the terms in the base language and, on each play, sends one of two signals. The metagame receiver then uses that signal to decide whether the current base-game signal has been reliable. Then the metagame agents reinforce if the metagame receiver is successful — that is, if she acts as if the past relative success rate of the current signal in the base game is *greater than or equal to* the specified threshold (0.75) precisely when it is. If successful, one term in the metagame will evolve to mean that the current base-game signal has in fact been at least as successful as the threshold and the other signal will evolve to mean that the current base-game signal has been less successful than the threshold.

On simulation, the base language typically (0.77) evolves to be as successful as possible given the random noise in the channel. When it does, one base-game signal *a* evolves to be more reliable than the threshold ($T = 0.75$) and the other *b* evolves to be less reliable. Further, a metalanguage that reliably (better than 0.90) sorts the base-game signals by their reliability above and below T typically (0.73) coevolves with the base language.

If both games are successful and if the base-game success rates are in fact stable, then one metalanguage term will mean that the probability that the current base-game signal will lead to a successful action is greater than or equal to the threshold, and hence *reliably true*, and the other metalanguage term will evolve to indicate that it is less than the threshold, and hence not reliably true.

But regardless of whether past current base-game success rates continue to hold, the metalanguage terms might be taken to represent the metagame receiver's degrees of belief in the truth of the base-game signal if she uses those terms as the basis for deciding whether or not to bet on the reliability of the current base-game signal at threshold odds. Further, for each successful run of the game at a different setting for T, a different likelihood distinction evolves and a different pair of terms to represent it. Together a family of such distinctions would allow for the evolution of increasingly fine-grained expressions of relative probabilities if the agents know which game they are playing. And, of course, the agents may also evolve a representation for *that*. This explains how it is possible for agents to evolve an increasingly fine-grained language for expressing their degrees of belief in the reliable truth of an expression.

When a stable base-language and a successful metalanguage evolves in this model, the terms of the metalanguage communicate the reliability of the base-game signals. More precisely, (1) if each base-game signal type comes to exhibit a stable reliability and (2) if the metalanguage coevolves to successfully indicate this reliability (and both of these conditions are typically met on simulation even just with simple reinforcement learning), then the metagame signals will coevolve with the base-game language to indicate whether each base-game signal in fact reliably communicates the current state of nature to the base-game receiver at the threshold level and hence leads to successful action for the base-game receiver. In this case, one of the metagame signal will indicate that the current base-game signal is *reliably true* and the other will indicate that it is not. A successfully evolved notion of reliable truth here, then, is an endogenous notion that tracks the past reliability of the base-game language and may also track the de facto reliability of the current base-game signal.

2.4 Understanding Truth in an Evolutionary Model

The metagame agents are rewarded for different types of action in the reflective truth and reliable truth models above. As a result, they evolve different pragmatic notions of truth. In the case of *reflective truth*, the inquirers learn to track and to communicate whether the descriptive conventions that coevolve in the base language are being honored on a particular play of the base-game. In the case of *reliable truth*, when the base-game exhibit stable reliabilities, the inquirers learn to communicate and to predict whether the description given in a particular play of the base game will lead to successful action.

That the base-game language and the metalanguage *coevolve* in these games is significant. Before there is a meaningful base-game language, there is nothing that could be true or false. In both models, the metagame evolves to track the truth of the base-game expressions even as those expressions themselves evolve meanings. A predicate that can be counted on to reliably indicate the truth of an expression requires that correspondingly sharp meanings have evolved. What is meant in taking a description to be true in each game coevolves with the significance of the base-game descriptions. On this account, the *meanings* of one's basic descriptions and the pragmatic significance of one's associated *notion of truth* become increasingly sharp together establishing a lasting link between meaning and truth on the particular notion.

The notions of truth that evolve in the games above are both extremely simple. For more subtle notions of truth to evolve, the agents would need to be playing correspondingly more subtle base games and metagames. A more sophisticated coevolved notion of truth might concern how expressions in the base language relate to the history of their successful use (e.g. in description, prediction, explanation, or argument), how they relate to each other (e.g. with respect to a specific variety of coherence), and/or how they relate to other aspects of the world to which the agents have at least indirect epistemic access (e.g. regarding success and failure in a particular variety of action under specified constraints).

A brief consideration of two logically distinct ways that a truth predict might be used in a formal setting illustrates this point. Truth predicates are sometimes used (1) to emulate higher-order quantification or (2) to represent infinite conjunctions. As an example of the first task, the expression "Everything the Pope says is true" might represent the higher-order quantified claim that "For all x, if the Pope says $\ulcorner x \urcorner$, then x". Such a notion of truth might evolve in the context of an evolutionary game where to be successful agents need to be able to communicate the recommendation to act on the advice of a Pope regardless of what advice he/she may ultimately give. This might happen if the Pope were always in fact reliable in his pronouncements. In contrast, an evolutionary game involving giving or evaluating mathematical proofs might require one to be able to represent and finitely express a concrete infinite conjunction. As an example, the finite expression "The theorems of Peano Arithmetic are true" might represent the single infinite conjunction "$(0=0) \wedge (0+1=1+0) \wedge \ldots$". In each case, the subtlety of the truth-telling task the agents face determines the subtlety of the game required to model the evolution of the corresponding notion of truth in that context. Any game involving the actual mathematical practice of giving and evaluating proofs, or any other nontrivial empirical, mathematical, or logical practice, would need to be significantly more sophisticated than anything we have considered here. But such an account would differ in complexity and subtlety, not in spirit.

In considering how a truth predicate for a basic descriptive language might coevolve with that language, we are considering how a metalanguage might coevolve with the language it describes. The metalanguage one gets depends on the evolution of the base language and on precisely how the agents are rewarded when the use that language.

The two truth games we considered in the earlier sections provide simple models for two pragmatic varieties of truth. In each case, the base language comes to allow for true descriptions in the relevant sense only as it comes to allow for successful coordinated action. In the first game, the sender attends to the coevolving *customary use* of expressions in the base language. This provides a weak coherence notion of truth, in some ways similar to Kuhn's, where proper use is common use. But in those cases where the base game evolves a successful signaling system, reflective truth in this model also tracks whether the base-game expressions in fact lead to reliable action and are in fact descriptive of nature. In the second game, the agents coevolve a base language and a notion of reliable truth that tracks whether its expressions *lead to successful action*. This provides a pragmatic notion of truth that is in some ways similar to Peirce's. But in those cases where the base game evolves signals with stable reliabilities and the metagame is successful, reliable truth in this model also tracks whether the base-game expressions are being used in the customary way and whether they are in fact descriptive of nature. An immediate moral is that an evolutionary game that rewards agents for tracking one sort of pragmatic truth may, under appropriate conditions, track other notions of truth as well.

The notions of *reflective truth* and *reliable truth* that evolve in these generalized signaling games are well characterized by the roles they come to play in each game. In this way, the games where these notions evolve and roles they come to play provide a natural means for characterizing and distinguishing between alternative pragmatic understandings of truth. And one would expect more subtle evolutionary games to pick out richer endogenous varieties of pragmatic truth.

Our best theories and descriptive language coevolve with the changing descriptive and explanatory demands of empirical science. We seek to describe the world but we also seek to provide reliable descriptions of our language and its proper use. When there is an incentive for doing so, inquirers may evolve a metalanguage that tracks various endogenous notions of truth. Such notions may coevolve with their descriptive language and theory to represent the evolving standards of inquiry. They are, like the rest of our coevolved language and theory, in the service of inquiry for the sake of what we take to be successful action.

3 The Evolution of Belief

Just as with truth, different evolutionary games provide different understandings of what it is to have and to represent a belief. Here we will consider a simple dispositional model of belief attribution. On this model, a descriptive language sufficiently rich to characterize an agent's beliefs might coevolve with that agent's basic linguistic dispositions. We will consider this in two steps. An agent (the metagame sender) might evolve a descriptive template to represent the dispositions of another agent (the base-game sender) in the context of a signaling game, then, by a process of template transfer, she may come to use the resulting language to characterize her own dispositions.

The base game here is a basic two-state/two-signal/two-act signaling game with unbiased nature as discussed earlier. This game is guaranteed to converge to a perfectly efficient signaling system on simple reinforcement learning.[8] We will suppose that it has evolved to be nearly optimal and that each term 0 and 1 has hence come to represent one of the two states of nature, either *A* or *B* (Fig. 6.4).

The metagame is also a two-state/two-signal/two-act signaling game with simple reinforcement learning, but the metagame sender tracks the base-game sender's dispositions to signal. What counts as the state of nature for the metagame sender is the contents of the base-game sender's urn that corresponds to the current state of nature in the base game. In particular, the metagame sender observes whether the base-game sender urn that corresponds to the current state of nature has more 0 balls or 1 balls. If it has more 0 balls, the metagame sender draws a ball at random from her more-0 urn; if it has more 1 balls, she draws from her more-1 urn. Suppose, for example, that the current state of nature in the base game is *B* and that the base game sender has more 0 balls in his *B* urn.

[8] This follows from the central result of Argiento et al. (2009).

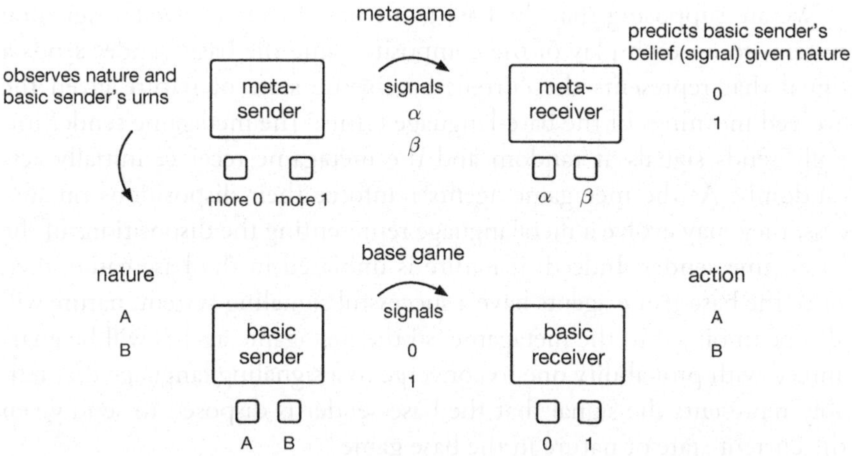

Fig. 6.4 Learning how to represent the dispositions of another agent (Source: Author)

The metagame sender would then draw from her more-0 urn and send the corresponding signal.

The metagame sender's urns each start with one α ball and one β ball. The metagame receiver has an α urn and a β urn. Each of these urns start with one 0 ball and one 1 ball. The ball she draws determines the metagame receiver's action on the current play of the metagame. The play is successful if the metagame receiver's action matches the signal that the base game sender sends on the current play of the base game. When successful, the metagame agents evolve a language where one term means that there are more 0 balls in the base-sender urn corresponding to the current state of nature and the other means that there are more 1 balls in the base-sender urn corresponding to the current state. Since the number of balls determines the relative likelihood of the base-sender's signals, when successful, the metagame agents will evolve a language that will allow the metagame receiver to reliably predict the base-game sender's signal given the current state of nature. In this way, the metagame signal may come to represent the base-game sender's disposition given the current state of nature.

We are supposing that the base game has already evolved a signaling system, so on each play of the composite game the basic sender sends a signal that represents the current base-game state of nature given the evolved meanings of the base-language terms. The metagame sender initially sends signals at random and the metagame receiver initially acts randomly. As the metagame agents reinforce their dispositions on success, they may evolve a metalanguage representing the dispositions of the base-game sender. Indeed, if nature is unbiased in the base game, then since the base-game agents have a successful signaling system, nature will also be unbiased in the metagame, so the metagame agents will be guaranteed with probability one to converge to a signaling language that reliably represents the signal that the base-sender is disposed to send given the current state of nature in the base game.[9]

The upshot is that if one supposes that dispositions track beliefs, then on a particular run of the composite game, one of the metagame terms α or β will evolve to represent "the base-game sender believes '0' " and the other will evolve to represent "the base-game sender believes '1' ". And, since the base game has evolved a signaling language, the base-game term 0 will represent either "the state of nature is *A*" or "the state of nature is *B*" and, whichever it represents, the base-game term 1 will represent the other. In this way, the metagame terms *come to represent the base-game sender's beliefs about nature* when the metagame evolves a successful language.

Of course, a metagame sender who understands the base-game language (i.e. one who has urns *A* and *B* with proportions of 0-balls and 1-balls that would allow her to communicate reliably with the base-game receiver) might observe *her own* dispositions given the current base-game state of nature. This is represented in the following figure. Here, when she signals either α or β to the metagame receiver, the metagame sender will be reliably reporting *her own belief* regarding how to describe the current state of nature in the evolved base-game language (Fig. 6.5).

Here the agents learn to represent their beliefs by learning to represent an agent's *dispositions to report* given the current state of nature. One might tell a similar story for how other notions of belief might evolve by

[9] This follows again from Argiento et al. (2009).

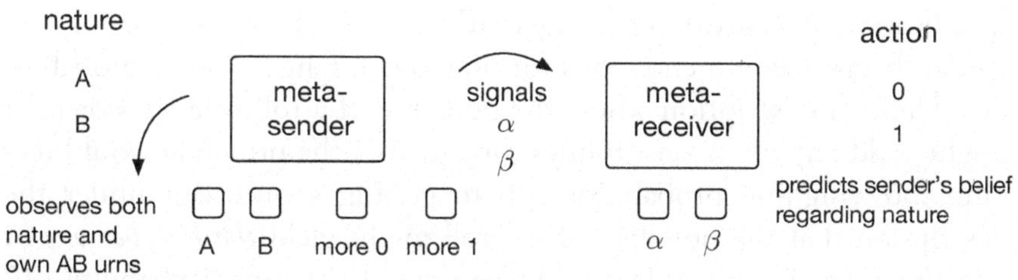

Fig. 6.5 Describing one's own dispositions (Source: Author)

considering alternative states of nature that the metagame agents might track. An evolutionary story involving the probabilistic distinctions discussed in the reliable truth section above might be used to explain how agents might evolve a language to characterize their *degrees of belief*. Similarly, if there were agents with epistemic access to intensions, say, and if they needed to be able to track and represent intensional states for the purpose of successful action, they might evolve a language that represents beliefs as intensional states. The point is just that, as with alternative notions of truth, one might individuate alternative notions of belief by considering those games where such meta-representations might evolve.

4 The Evolution of Indifference

Evolutionary games can also provide insight into proposed epistemic principles such as the principle of indifference. The principle of indifference says that, in the absence of other information, one should assign probability $1/n$ to each of n possibilities. If one has a die with six sides, for example, the principle would recommend assigning a probability of $1/6$ to each possible outcome. The philosophical question is whether, when, and to what extent such a recommendation might be rational.[10]

As a general principle of reason, there is little to recommend the principle of indifference. It faces a number of significant conceptual problems. Indeed, the principle itself is not even meaningful without the

[10] See Zabell (2016) for a brief introduction to the history of the principle of indifference with a particular focus on von Kries' worries regarding the principle.

specification of a partition over alternatives since different partitions provide different recommendations for how one ought to assign probabilities. Under a description where the outcome of a roll of a six-sided die might yield any of six possibilities (one to six), the principle would recommend assigning probability 1/6 to getting a one. But under the description that the outcome of the roll might yield *the number one* or *something else*, the principle would recommend assigning probability 1/2 to rolling *the number one* and probability 1/2 to rolling *something else*. One might be tempted to object that this partition of outcomes is somehow unnatural. But the naturalness of a partition depends on context.

Bas van Fraassen's (1989) cube factory story illustrates the point. Suppose one only knows that a factory produces cubes with a side between 0 and 2 meters. If one considers side-length, one might imagine that the principle of indifference requires that one take the probability of a randomly selected cube having a side between 0 and 1 to be 1/2 since side-lengths range from 0 to 2. If one considers volume, one might imagine that the principle of indifference requires that one take the probability of a randomly selected cube having a volume between 0 and 1 to be 1/8 since volumes range from 0 to 8. But since having a side length between 0 and 1 is the same thing as having a volume between 0 and 1, the different partitions *each perfectly natural given different interests* yield inconsistent probability assignments.

The upshot is that (1) there is no content to the principle of indifference without a specified partition since different partitions yield inconsistent probability assignments and (2) the partition one finds most natural will depend on one's interests. And this puts us back where we started. Inasmuch as the principle of indifference must be tailored to a specific application, any particular proposed principle and partition is a poor candidate for a general principle of reason.

There is a further problem. If the principle of indifference were in fact a basic principle of reason, one would expect that something would go wrong if an agent were *not* to use it. It is entirely unclear, however, what such bad consequences might be. A good Bayesian might assign any set of coherent, non-dogmatic priors (that is, probabilities that satisfy the standard axioms of probability theory and are neither zero or one) to her *n* hypotheses without fear of finding herself committed to a Dutch Book or failing to respond appropriately to relevant evidence. And, as she conditions on new evidence, she will expect that her degrees of belief will reflect relative frequencies and hence chances in nature.

All this said, however, it is *possible* for a language to evolve so that the terms of that language partition nature in such a way that the principle of indifference assigns probabilities over the partition that are in fact approximate equal to the relative frequency of each type of state in nature. Indeed, this might happen under relatively plausible, generic conditions. Here we will consider one way that this might happen.

Consider a signaling game where there are many more states of nature and corresponding acts than there are terms that the sender might use to signal the receiver. In this case, the agents lack the expressive resources to evolve a term language that perfectly communicates the current state of nature. At best, the agents will evolve a signaling system where multiple states of nature correspond to the same term. Hence, the evolved language will never be perfectly successful. But it might be more or less optimal given the agents' expressive constraints.

When there are more states than terms, it can happen that a language evolves in such a way that each of several states of nature trigger the sender to send a particular corresponding term. In this way the evolved meanings of the terms may come to partition the states of nature with each term denoting an element of the partition containing the states that correspond to that term.

The most efficient signaling system the agents might evolve would be one that communicates the most information about the current state of nature per signal. The Shannon entropy over a partition is given by

$$H = E[I(i)] = \sum_i -p_i \log_2 p_i$$

where p_i is the probability associated element with element i of the partition. This expression is maximum when the probabilities corresponding to each element of the partition are equal. The expected information $E[I(i)]$ (in bits here) of a signal that indicates element i is also maximum when the probabilities are uniform. The idea is that a signal communicates the most information possible when one is maximally uncertain regarding what that signal will be. So the agents' evolved language communicates the most information per signal if the probability of each type of signal is equal, and this occurs if and only if the signals partition the states of nature into equally probable sets.

Since a language where the terms partition nature into equally probable sets allows the agents to communicate the most information per signal, one might expect such a language to evolve contexts where agents are somehow rewarded for efficient communication. This might happen if the agents must pay a cost for each signal they send. If a maximally efficient language evolves, it will induce a partition over the possible states of nature such that a principle of indifference over that partition will yield probabilities that in fact agree with the past relative frequencies of each element in the partition. Here the success of agents employing the principle of indifference to set their priors would be an artifact of the very special evolutionary context in which their language evolved, a context that rewarded efficient signals.

Languages that induce an unbiased partition may, however, evolve even without the efficiency constraints imposed by costly signaling. Indeed, languages that partition states into equiprobable types often arise from simple reinforcement learning alone. In order to get an idea of how this might occur, we will turn to a concrete example.

Consider a signaling game with ten equiprobable states of nature, two terms, and ten acts, one corresponding to each of the states. Suppose that both the sender and receiver learn by simple reinforcement. Since each state of nature is equiprobable, an evolved language on a $10 \times 2 \times 10$ signaling game will partition the states into equiprobable sets if and only if each term comes to be triggered by precisely five types of state. We will represent this partition as (5, 5) and the other possible partitions similarly (Fig. 6.6).

On 1000 runs of the $10 \times 2 \times 10$ signaling game with simple reinforcement with 10^6 plays/run, equiprobable partitions of nature or partitions that are nearly so are significantly more likely than not. The evolved language induces a (5, 5) partition on the states of nature about 0.31 of the time, a partition (6, 4) where the first term *A* comes to represent six states and the second term *B* comes to represent four states occurs about 0.24 of the time, and a partition (4, 6) where the first term *A* comes to

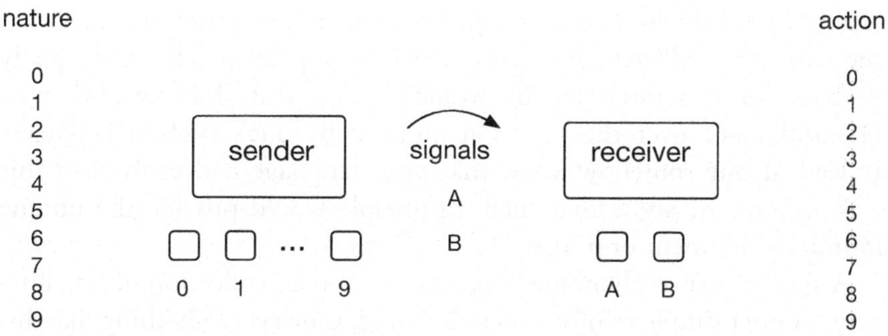

Fig. 6.6 Evolving an unbiased partition (Source: Author)

Table 6.1 Evolved partitions for the two terms (term A, term B) over 1000 runs with 10^6 plays per run (Source: Author)

Partition	Relative frequency
(5, 5)	0.310
(6, 4)	0.238
(4, 6)	0.244
(7, 3)	0.073
(3, 7)	0.075
(8, 2)	0.019
(2, 8)	0.009
(1, 9)	0.000
(9, 1)	0.001
(10, 0)	0.000
(0, 10)	0.000

represent four states and the second term *B* comes to represent six states occurs about 0.24 of the time. The evolved language, then, induces an equiprobable or nearly equiprobable partition over the possible states of nature about 0.80 of the time. See the table below for further details. Note that the likelihood of the less even partitions drops off very quickly (Table 6.1).[11]

Given the symmetry of the game, one might expect each of the two evolved terms to obtain about half of the time. But the simulation results here suggest something significantly stronger. If nature continues to

[11] The simulations in this section were run by Travis LaCroix as part of our joint research regarding conditions under which unbiased partitions might evolve.

behave as it did while the language was evolving, one should expect that the two terms will *partition* nature, at least approximately, into equally probable sets of states. Here one would be successful adopting a principle of indifference over the partition induced by one's evolved language. Indeed, if one somehow knew that one's language had evolved in this way, one might argue that such a principle would provide the unique rational assignment of priors.[12]

A similar sort of phenomena occurs for a broad collection of signaling games under simple reinforcement learning. One gets something like this in the $9 \times 3 \times 9$ signaling game where three terms evolve to partition nine equally-probable states of nature. Here, on simulation, the evolved partition is typically an even (3, 3, 3) or something very close to it like (2, 3, 4).[13]

One also sees this phenomena with simple reinforcement learning when the states of nature are themselves not equally probable. Consider a $4 \times 2 \times 4$ signaling game where one state occurs with probability $1/2$ and the other three states occur with probability $1/6$ each. In this case, agents learning by simple reinforcement typically end up associating one of the two available terms with the most likely state and the other term with the three less likely states. The upshot is that the two evolved terms induce an equiprobable partition on nature, and here with a different number of distinguishable states associated with each term. In each of the three cases here, then, simple reinforcement learning evolves terms that end up referring to "natural kinds" that are roughly equiprobable.[14]

But it would be a mistake to imagine that there is a general argument for the principle of indifference here. While it might work in very special contexts of costly signaling or simple reinforcement learning, it will not work very broadly. If we change the learning dynamics or the payoffs, all bets are off. Regarding the importance of payoffs here, simple reinforcement learning should only be expected to induce an equiprobable partition of nature *for unbiased payoffs*. If one state of nature is particularly

[12] Of course, if one had evidence that one's language had evolved in this way, this would be more honestly characterized as an assignment of probabilities inferred from posteriors than as a basic assignment of priors.

[13] See Barrett and LaCroix (2020) and LaCroix (2020) for extended discussions and the empirical details of such games.

[14] See Barrett (2007) for a basic account of the evolution of natural kinds.

important to get right for successful action, that state might well get its own term even if it is relative improbable. All considered, it is entirely unsurprising that the principle of indifference typically simply fails to work even on partitions that seem natural given our actual evolved languages and interests.

5 Discussion

If one believes that our natural descriptive language has coevolved with our knowledge of the world and our deliberative practice, then one is committed to there being an evolutionary story that explains our descriptive language, knowledge, and deliberative practice, and how they work together in the context of inquiry. Our actual evolutionary story will be extremely complex with many of the details lost in the history of our epistemic ancestors' successes and failures in inquiry. But this does not mean that we can say nothing about how we ended up with our current epistemological practice. If a particular feature of our descriptive and predictive practice is sufficiently simple, one might be able to find a correspondingly simple and generic evolutionary model.

In this spirit, we have considered how generalized signaling games might be used to model various pragmatic notions of truth, how such notions might coevolve with a descriptive language, how inquirers might evolve a language that allows them to characterize their beliefs and degrees of belief, and evolutionary conditions under which one's language might evolve to satisfy a principle of indifference over expressible states of nature. Each of these models has been as simple as possible given the phenomena to be explained — in some ways arguably too simple. It is worth reflecting on sort of explanations such models might provide.

The evolutionary models we have considered explain how a number of salient features of reflective inquiry might have evolved from the activities of agents with only meager cognitive resources, resources so meager they are perhaps better referred to as basic *dispositions*. The modeled inquirers here begin with no descriptive language or special inferential abilities. They just have the ability to update how they interact with each other and the world based on the results of earlier interactions by reinforcement on success. From this alone, they evolve the basic capacity to describe the world they inhabit, successfully predict the future on the basis of their descriptions, and characterize their own beliefs and commitments.

These models show *how it is possible* for such rudimentary capacities to coevolve, but they do more than that. They show *how agents might evolve such capacities starting with only simple, generic dispositional resources.* Reinforcement learning is exceedingly basic. It requires only a mechanism by which one type of event might become more likely the more often another type of event has occurred. Such processes are ubiquitous in nature. They are routinely exhibited by even inanimate systems. The more often runoff from rain follows a particular trajectory down a desert ravine, the more likely future runoff will follow a similar trajectory.

If the behavior to be explained is manifestly useful to the agents exhibiting it and if the evolutionary story one tells is particularly robust and if the resources required are sufficiently generic, one might even be able to explain why such an evolution *should be expected* under a broad range of conditions. Such is the case with signaling and simple inductive prediction. Both are in fact common in nature at every level of life. This is in part explained by how easily they might evolve in the context of exceedingly simple evolutionary games under a generic learning dynamics like simple reinforcement.

Whenever the required dispositional resources are in place, and random physical processes together with enough time suffices to ensure this, we will see the evolution of organisms with something very like our cognitive faculties, including the ability to represent the world and to use descriptive language for the purpose of successful cooperative action. This does not mean that such faculties have always been commonplace nor does it mean that they will persist once they have evolved. But it does begin to explain how empirical inquiry might be the result of simple evolutionary processes like those we have considered here all the way down. Of course, that is what the confirmed naturist always believed in her heart, but there is understanding to be gained by attending to the details.

Each of the stories here is a small part of the larger story of accounting for how agents might evolve a useful descriptive language then use that language for reflective inquiry. Together, they illustrate the explanatory reach of applying evolutionary game theory to epistemology. Agents who start with no special knowledge, concepts or rational faculties may evolve a basic descriptive language. They may coevolve predictive dispositions that are closely coordinated with their descriptive language. Their language may be sufficiently expressive to describe their own dispositions and hence represent their beliefs and commitments. And their descriptive language may continue to coevolve with the knowledge it makes possible.

References

Argiento, R., Pemantle, R., Skyrms, B., & Volkov, S. (2009). Learning to signal: analysis of a micro-level reinforcement model. *Stochastic Processes and Their Applications, 119*(2), 373–390.

Barrett, J. A. (2007). Dynamic partitioning and the conventionality of kinds. *Philosophy of Science, 74*, 527–546.

Barrett, J. A. (2008). Approximate truth and descriptive nesting. *Erkenntnis, 68*(2), 213–224.

Barrett, J. A. (2016). On the evolution of truth. *Erkenntnis, 81*(6), 1323–1332.

Barrett, J. A. (2017). Truth and probability in evolutionary games. *Journal of Experimental and Theoretical Artificial Intelligence, 28*(1), 219–225.

Barrett, J. A. (2021). Scientific Inquiry and the Evolution of Language. In W. J. Gonzalez (Ed.), *Language and Scientific Research* (pp.121–147). Cham: Palgrave Macmillan.

Barrett, J. A. (forthcoming). Self-assembling games and the evolution of salience. *British Journal for the Philosophy of Science.*

Barrett, J. A., & LaCroix, T. (2020). Epistemology and the structure of language. *Erkenntnis* (Forthcoming). https://doi.org/10.1007/s10670-020-00225-4.

Barrett, J. A., & Skyrms, B. (2017). Self-assembling games. *The British Journal for the Philosophy of Science, 68*(2), 329–353.

Barrett, J. A., Skyrms, B., & Mohseni, A. (2018). Self-assembling networks. *The British Journal for the Philosophy of Science, 85*(5), 910–920. https://doi.org/10.1093/bjps/axx039.

Barrett, J. A., Skyrms, B., & Cochran, C. (2020). On the evolution of compositional language. *Philosophy of Science, 85*(5), 910–920.

Houser, N., & Kloesel, C. (eds.) (1992). *The Essential Peirce: Selected Philosophical Writings Volume 1 (1867–1893).* Bloomington and Indianapolis: Indiana University Press.
Kuhn, T. S. (1996). *The Structure of Scientific Revolutions.* Chicago: University of Chicago Press.
LaCroix, T. (2020). Communicative bottlenecks lead to maximal information transfer. *Journal of Experimental and Theoretical Artificial Intelligence* (Forthcoming). https://doi.org/10.1080/0952813X.2020.1716857.
Lewis, D. (1969) *Convention.* Cambridge, MA: Harvard University Press.
Skyrms, B. (2010) *Signals: Evolution, Learning, & Information.* New York: Oxford University Press.
van Fraassen, B. (1989) *Laws and Symmetry.* Oxford: Clarendon Press.
Zabell, S. (2016). Johannes von Kriess *Principien*: a brief guide for the perplexed. *Journal for General Philosophy of Science, 47*, 131–150.

7

Models, Fictions and Artifacts

Tarja Knuuttila

1 Models, Representation and Languages of Science

Modeling is certainly the scientific practice that relies on the richest repertoire of diverse languages and representational vehicles: mathematical, computational, diagrammatic, natural spoken and written languages, and concrete objects of various kinds, spanning from material and digital 3D objects to various biological preparations. Moreover, the recent philosophical interest in representation has surfaced precisely in the context of modeling. One would thus have expected the discussion of representation to address the various representational languages and other vehicles of model construction, but this has not been the case. The discussants have, instead, focused on the relationship of models to worldly target systems in a general and formal manner, the main issue being whether representation can be understood in terms of similarity, isomorphism or some other kind of morphism (e.g. Suárez 2003; Frigg 2010; Bueno and Colyvan 2011; Chakravartty 2010). To be sure, especially the

T. Knuuttila (✉)
Department of Philosophy, University of Vienna, Wien, Austria
e-mail: tarja.knuuttila@univie.ac.ai

W. J. Gonzalez (ed.), *Language and Scientific Research*,
https://doi.org/10.1007/978-3-030-60537-7_7

complexities of mathematical languages and their suitability for representing the world has been addressed in the discussion of representation. Yet even in these discussions the main goal has still been that of arguing for (or against) the structuralist/semantic conception that approaches representation through mappings between a model and a target system (cf. Bueno and French 2011; Pincock 2004).

In addressing the roles different representational languages and tools play in modeling, I will leave the aforementioned discussion of scientific representation aside, concentrating instead on two novel approaches that aim to give a more fully-blown account of modeling. The first one of them, the DEKI account (e.g. Frigg and Nguyen 2016, 2017), still attempts to tackle the problem of model-based representation head-on, but it also recognizes the different kinds of models, aiming to treat both concrete and nonconcrete models in a unified manner. In contrast, the other approach, the artifactual account, seeks to bypass the problem of accounting for the representational relationship, focusing on the epistemic uses of various representational tools.

I will argue that despite its unificatory aims, the DEKI-account treats nonconcrete and concrete models non-symmetrically. It supposes that in the case of concrete models — e.g. scale models and other physical models — the physical object is the model, whereas in the case of nonconcrete models — e.g. mathematical models — the model is an imagined-object rendered with a model description. The problem with this account is that in distinguishing imagined systems from the model descriptions conveying them, Frigg and Nguyen leave unexplained how scientists' imaginings are related to the enablings of different kinds of representational tools, or coordinated among different scientists, and with real-world phenomena.

I will propose an artifactual alternative that treats models as purposefully created epistemic tools that are constructed for the study of certain scientific problems. The artifactual account places languages of science on the center stage. It addresses the actual representational tools that are used in model construction, highlighting their epistemic functioning: how they enable, shape and delimit scientific reasoning and imagination. The artifactual account is able to give a unified account of the wide variety of different kinds of models; it treats them as entities with both abstract and concrete dimensions, realized in different representational modes and media.

In what follows, I will first discuss Roman Frigg's and James Nguyen's DEKI account of representation (Sect. 2) and its philosophical critique (Sect. 3). I will then introduce the artifactual account of models (Sect. 4).

2 The DEKI Account of Model-Based Representation

Frigg and Nguyen's DEKI account is an elaborate construction weaving Nelson Goodman's (1976) and Catherine Elgin's (e.g. 2004) theory of pictorial representation together with Kendall Walton's theory of make-believe (1990). Frigg and Nguyen develop the DEKI-account from a case of a concrete vehicle. Following the philosophical tradition, their prototypical model is a three-dimensional object with some distinct properties. Frigg and Nguyen call this object a "base," or an "O-object," where "O" specifies what kind of a thing an object is. In line with the pragmatic view of representation, such an object becomes a representation first when an agent chooses to *use* it to represent something. Central for Frigg and Nguyen's account is the idea of *representation-as* that they derive from Goodman and Elgin. An object may be used to represent something as something else, thereby becoming a Z-representation (e.g. Elgin 2010). For instance, there is a genre of caricatures representing Winston Churchill as a bulldog. In this case, an O-object, a picture of Churchill, is interpreted as a Z-representation (a bulldog-representation) becoming thereby a vehicle of representation.

The DEKI-account of representation refers to the notions of *denotation, exemplification, imputation*, and *keying* up. In order to represent, a vehicle has to both denote a target system and exemplify some properties, *imputing* those properties, or related ones, to a target system. However, the properties exemplified by a model cannot usually be directly imputed to some target as the model and the target are usually very different kinds of things. A *key* is needed to translate the properties exemplified by the representational vehicle into properties of a target system. Frigg and Nguyen

(2018) analyze Phillips-Newlyn hydraulic model of macroeconomy that is a concrete object consisting of pipes and reservoirs with water flowing through it. This example shows that a material machine can, under interpretation, exemplify some features to be imputed to actual economies, e.g., stocks and flows of commodities (cf. Frigg and Nguyen 2018).

In the case of mathematical models, there seems not to be any concrete object doing the representing, and so Frigg and Nguyen treat them as fictional constructs, imagined-objecs, presented by model descriptions:

> Nonconcrete models are typically presented through descriptions, portraying things like spherical planets and immortal rabbits. We call these descriptions *model descriptions*. This gives us the essential clue: model descriptions are like the text of a novel: they are props in games of make-believe. (Frigg and Nguyen 2016, 237)

The notion of a prop and make-believe are adopted from Kendall Walton's theory of mimesis as make-believe (cf. Walton 1990). One attractive feature of Waltons' account is its solution to the problem concerning the ontological status of fictional entities: there are no such entities, only pretense. From this perspective, fictional stories merely *pretend* to assert something about real people and places. The fictional characters, and the literary works they inhabit, are just props in the game of make-believe. And this also applies to other kinds of make-believe. Any thing that can affect our senses and is furthermore subject to "a principle of generation" can function as a prop in a game of make-believe — like toys in children's games (cf. Walton 1990, 38).

In scientific modeling, it is *model descriptions* that function as props that prompt scientists to imagine a model system. The model system is an imagined-object that due to the principles of generation "can have properties that have not been written into the original model description" (Frigg and Nguyen 2016, 237). Consequently, the imagined-objects inhabiting scientists' minds are, according to Frigg and Nguyen, richer than what can be read from the model descriptions alone. Such richness derives from the way the model description, combined with principles of generation, creates an imagined-object. The imagined object can then have some pretended properties, i.e. it is fictional in the model that the imagined-object has such properties. Imagination is not free from constraints, however. The game of make-believe constrains imagination due to facts about the props and the principles of generation. Frigg and Nguyen do not say too much about how this is supposed to happen, they merely mention that principles of generation involve "background theories" (2016, 11). About model descriptions they also have little to say but notice that "mathematics can enter models in two places: in the model description and in the rules of generation" (15). These sparse remarks are not that surprising given that Frigg and Nguyen's focus is on the imagined-object, which they consider as the vehicle of representation: "By mandating those involved in a certain game to imagine certain things, the model description generates the imagined-object that serves as the vehicle X of a representation as." (2016, 13). But the critics have been skeptical of positing such intermediate fictional entities (e.g., Toon 2012; Weisberg 2013; Levy 2015).

3 Some Problems of Imagined-Objects

Levy (2015) and Toon (2012) argue against the fictionalist detour through unobservable imagined objects. Levy argues that such objects "by virtue of their non-actuality, […] are not the kinds of things we can observe and come into contact with" (784–785).[1] Instead of such indirect approaches Toon and Levy prefer what they call "direct approaches to representation." Somewhat surprisingly, however, they are also making use of Walton's make-believe in order to account for the idealized and simplified nature of scientific models. They consider models as "imaginative descriptions of real-world phenomena" (Levy 2015, 797) that "prescribe us to imagine things about the actual system" (Toon 2010, 308). But this approach has difficulties with models that do not have real world-targets, or models with general targets (see, e.g., Frigg and Nguyen 2017, 109–112). Perhaps the biggest problem of the direct make-believe construal is its inability to account for how models enable surrogative

[1] He levels the same kind of critique also against the approaches that take models to be abstract objects (Weisberg 2013; Giere 1988).

reasoning. This seems an unnecessary sacrifice as surrogative reasoning is generally considered as central for many modeling practices. But Toon's and Levy's criticism does not actually concern indirect approaches per se, but rather the way they postulate fictional (or abstract) entities that lack sensuous qualities. The question is how can imagined-objects — being mental phenomena (see Frigg 2010), and as such not intersubjectively available — act as vehicles of surrogative reasoning in such a collective endeavor as science. At least three issues seem to appear.

First, if fictions are analyzed in terms of the pretense theory, it becomes a problem how these imagined-objects are related to real-world objects and systems. Since an imagined-object is strictly speaking non-existent, its features are uninstantiated that makes any comparisons to the real world objects difficult. It is not clear why the uninstantiated properties of imagined objects should be less problematic than those objects themselves (whose ontological status the pretense account was designed to solve).[2] It is difficult enough to explain how external representations can stand for and give knowledge about real-world systems, and so to ask the same question about the imaginings of the scientists seems even more challenging. Moreover, this question leads us from philosophy of science to the realm of cognitive science — and the thorny issues of mental representation. Frigg and Nguyen do not try to solve this problem. They simply mention that the DEKI account of representation does not "require comparative claims" and refer to Salis' (2016) proposal of how model-world comparisons could be made.

Second, the problem concerning the relationship between model descriptions and imagined-objects is no less taxing, especially as Frigg and Nguyen (2016, 2017) do not explain how model descriptions combined with principles of generation generate richer imaginary worlds in the minds of scientists. Presumably, they think that this cannot be a problem philosophy of science needs to address, since it is an everyday phenomenon in how literature is being received. Yet the problem is central for philosophy of science, as the changes in the mathematical representation of a model frequently alter many of its epistemically relevant properties. Weisberg (2013) highlights this shortcoming of the fictional account.

[2] See Godfrey-Smith 2009 for discussion.

A case in point is provided by his analysis of the Lotka–Volterra model, and what difference it makes when the model is presented in individual-based methods in contrast to ordinary differential equations (cf. Weisberg and Reisman 2008). A fictionalist might want to claim that the two different mathematical representations describe the same imagined system, especially as one of the benefits of the fictional approach is that of maintaining the identity of the model system under different descriptions (e.g. Frigg 2010, 256; Frigg and Nguyen 2016, 231). Yet, if it is the fictional system that is supposed to represent, or resemble, the target system, the epistemic consequences of using different, mathematical, or computational modeling methods are left without recognition.

Finally, there is the problem of how the imaginings of different scientists are supposed to be coordinated. In his criticism of the fictional accounts, Weisberg (2013) also addresses the variation in the imaginings of individual scientists. Even if the problem of comparing the features of imaginary systems to the real-world systems were solved, how can we ascertain that we are dealing with the features of the same imagined-objects. Frigg suggest that "[a]s long as the rules [of a particular game of make-believe] are respected, everybody involved in the game has the same imaginings" (2010, 264). But this solution does not recognize that what *is* intersubjectively available to scientists are the representational tools (i.e. "model descriptions") with which models are constructed and whose properties, in the modeled configuration, modelers are studying. The rules and norms concern the use and interpretation of these external representations, in particular scientific contexts, in which the relevant background knowledge is embedded.

The question is, what is the philosophical added value of invoking additional imagined-objects? What kind of rules of generation, over and above the ones concerning the use of shared external representational means, in some specific contexts, are needed to understand modeling practices? Moreover, it does seem that the ascent from model descriptions to imagined-objects tends to set aside important aspects of scientific work. For instance, modelers pay a lot of attention to the particular mathematical abstractions they are using, and the modeling choices due to approximation and tractability. Yet these choices and their justification tend to get ignored if the focus is on pretend properties of imagined-objects. And what about model systems, whose workings cannot be imagined or understood? How are we to ascribe fictional features to computational models, whose computational processes are quite opaque to the human mind? It is difficult if not impossible to mentally simulate the dynamic of interactions in non-linear complex systems, or imagine probabilistic or high-dimensional models. This is precisely one of the most important reasons for why these models are so indispensable tools for scientific practice.

The aforementioned problems are due to the way the fictional account of modeling sets apart model descriptions and imagined-objects, locating the most important epistemic role in the latter.[3] While the question of representation is already a difficult one to tackle on its own, the reliance of the DEKI account on imagined-objects in the case of nonconcrete models makes the question even more complex. Moreover, making imagined-objects the locus of representation largely ignores how humans are able to creatively extend their cognitive capabilities by developing and using an ever expanding and diversifying array of representational and computational tools. The cognitive sciences have accumulated abundant evidence on how different representational tools are crucial for our cognitive accomplishments, studied within diverse approaches such as extended, distributed and enactive cognition (e.g. Clark 2008; Hutchins 1995). Zhang (1997) presented an important early experimental study on how particular representational devices facilitate reasoning. He showed that the same abstract structures conveyed by different representational devices have different enablings for human reasoning. Vorms (2011) argues essentially for the same point using scientific examples. She demonstrates how theoretical representations that are identical from the formal and empirical point of view can nevertheless expedite different kinds of inferences.

[3] Godfrey-Smith (2006, 2009) has also put forth a fictional account that relies on the distinction between model systems and model descriptions. Knuuttila (2017) and Salis (2019) have recently argued, though approaching scientific models from different perspectives, that model descriptions should be considered as parts of models.

To conclude, the problem of the DEKI account when it comes to "nonconcrete models" is not due to its indirectness but rather its reliance on the supposed representational abilities of imagined-objects. To understand the contemporary modeling practice, it seems important to make room for surrogate reasoning and also for the modal nature of modeling. One of the main motivations for engaging in modeling is the exploration of various possibilities, and hypothetical or imaginary scenarios. In the following, I will present an artifactual alternative to the DEKI-account that can, as well, accommodate surrogate reasoning, and address the modal dimension of modeling.

The artifactual approach differs from the representational accounts of modeling in not supposing that models give us knowledge in virtue of representing some actual target systems more or less accurately. However, the artifactual approach is not incompatible with the DEKI account of representation, apart from two important respects. First, the artifactual approach does not divide models into concrete and nonconcrete models, the latter having no objective existence of their own (apart from the imaginings of the scientists). Instead, the artifactual account also treats mathematical models as objects that are susceptible to manipulation, although the representational media used plays a different epistemic role in mathematical modeling than, for example, in scale modeling. Second, in considering scientific models as concretely embedded intersubjectively available objects, the artifactual approach does not set apart "model descriptions" from the (fictional or abstract) content they convey. The representational tools used in model construction are envisaged as inseparable ingredients of a model.

4 The Artifactual Account of Models

From the artifactual perspective, models are like any other artifacts in that they are human-made or altered objects intentionally produced and used for some purposes within the sphere of particular human activities (e.g. Knuuttila 2005, 2011, 2017). They are concretely constructed things, making use of various representational tools and material media. As artifacts they are constructed for certain purposes, although they may also be repurposed for other uses. In scientific research, such purposes are many: explanation, exploration, prediction, experimental design, didactic uses etc. Of these uses, the philosophy of science discussion has mainly been interested in the epistemic ones, supposing that in order to give us knowledge, or to explain, models need to be able to represent their actual target systems. While this assumption has often seemed too obvious to even be questioned, giving an account of representation has proven difficult (e.g., French and Ladyman 1999; Suárez 2003; Frigg 2010).

The DEKI account is perhaps the most developed — and, to be sure, the most convoluted — attempt to analyze scientific representation. Apart from the problems concerning imagined-objects, it is plagued by the same problem as many other accounts of representation, except for the most deflationary ones: it assumes too much to be known. Representational relation is accomplished by imputing the properties of the model system to the target system according to a key. But where does the key come from, and how are we to articulate the target such that keying up becomes possible in the first place? I do not wish to contest the possibility of keying up per se, but simply to point out that for such a translation from the properties of the model to the features of the target system to succeed, a lot of epistemic work has already been done. The question is how this is accomplished. It seems that the activity of modeling plays an important part in such exploration, and so a clue to the epistemic functioning of models can be found in the processes of constructing and using them.[4]

The volume *Models as Mediators* (Morgan and Morrison 1999) was crucial in shifting the philosophical interest in models towards the practice of modeling. The focal point of the analysis of Morrison and Morgan (1999) is on how scientists *learn* by building and manipulating models. This constructive and interventional side of modeling is central for the artifactual approach. It highlights how scientists gain knowledge through articulating and working with the different relationships built into the model — instead of supposing that *the* route to knowledge is through (at least) partially accurate reproduction of the actual state of affairs in the world. In contrast, models as external artifacts allow epistemic access to many theoretical and empirical problems by enabling various inferences

[4] For explorative modeling, see Gelfert 2016.

(Suárez 2004), providing new results, and, as a consequence, possessing considerable modal reach (cf. Godfrey-Smith 2006). The important thing to note is that for a model to be manipulable and experimentable, and able to be communicated, it has to have an intersubjectively available concrete embodiment.

For understanding the epistemic functioning of modeling, then, the concrete embodiment of a model needs to be considered as an integral part of it, and not just a "model description." The artifactual account approaches this concrete dimension of modeling through the distinction between representational mode and media (e.g. Kress and van Leeuwen 2001). The epistemic importance of material embodiment is obvious in the case of physical models but it also applies to mathematical and other "nonconcrete" or abstract models. The point is that depending on the type of a model, its concrete implementation by available representational tools and materials plays different epistemic roles. This is what the distinction between representational modes and media aims to pinpoint.

4.1 Representational Modes and Media

One of the most conspicuous features of the contemporary modeling practice is the amazing variety of representational tools it makes use of. It is somewhat ironic that in its fixation on the representational relation between a model and its target system, the representational approach does not consider the actual representational devices that scientists use in model construction. To get a firmer grip on the epistemic functioning of different kinds of representational devices, representational mode needs to be distinguished from representational media. The *representational mode* refers to the different symbolic or semiotic devices (pictorial, linguistic, mathematical, diagrammatic etc.) with which various meanings or contents can be expressed. Such representational modes are embedded in *representational media* that encompasses the material means with which representations are produced (such as ink in paper, a digital computer, biological substrata and so forth). For instance, natural language is a representational mode that can be realized by different media, either as speech or as writing (cf. Knuuttila 2011). In model construction and use, representational mode and media are often closely coupled, yet it is analytically useful to distinguish between them. The distinction enables a more unified treatment of different kinds of models. For example, the material embodiment of mathematical models is crucial for their manipulation, yet the concrete media plays a different, more prominent role in physical three-dimensional models than in mathematical models. Below, I will briefly discuss the representational modes and media in some different model types, ordered according to the degree of the importance of their concrete material dimension. It is important to note that all model types have both abstract/ "nonconcrete" and concrete dimensions, though in varying combinations.

(i) Mathematical Models

Highly idealized mathematical models provide the prototype for the fictional models in science. It appears less natural to speak about fiction in relation to diagrams or scale models, since they make use of the iconic, geometrical and material properties of the representational tools in question. Yet on the artifactual account there is no in principle difference between these different model types. It is just that from the point of view of scientific practice, representational modes and representational media, respectively, play different roles in their scientific uses (the representational mode and the representational media providing the two entangled dimensions of any representational devices). In the case of mathematical modeling, the representational media is less important and the focus is on the representational mode. For instance, in mathematical modeling of genetic networks, one can use different methods such as coupled ordinary differential equations (ODE), Boolean networks, and stochastic methods, all representing different mathematical representational modes. As for the representational media, it does not (usually) matter whether the media is chalkboard and chalk or whiteboard and erasable markers, for example. But the external scaffolding they provide for mathematical reasoning and imagination is nevertheless crucial for memorizing, manipulation, computing and demonstration.

(ii) Computer Simulations

The step from a mathematical model, e.g., ODEs, to a simulation model serves as an example of a change in representational mode. As the recent philosophical discussion of simulation has stressed, the discretized programmed equations of the simulation model stand in no straightforward relationship to the equations of the basic theoretical model. An illustrative example of a change of representational mode is presented by Weisberg and Reisman (2008), who discuss the reconstruction of the Lotka–Volterra model (i.e., population-level ODE model) in the individual-based framework. This change of representational mode has epistemic consequences: population level models are simple and more tractable, but they do not provide the means to study local-level interactions between individual organisms.

As for the representational medium, there has been philosophical discussion of what role a material artifact, the digital computer, plays in simulation (see Parker 2009). Can simulations be considered basically immaterial or not? The majority of the discussants agree that the fact that computer simulations are implemented on a concrete device and thus involve physical processes, when run on it, provides them with a material status. But then the next question concerns the epistemic role of digital computers. Should we regard them as mere computational aids, as number-crunching devices, or does computational modeling introduce epistemic characteristics of its own? Humphreys (2009) is among those who claim that it does: there are several features of simulation that distinguishes it from mathematical modeling, such as epistemic opacity and the dynamic temporal nature of computation.

(iii) 3-D Physical Models

Consider two three-dimensional physical models in both of which the flow of water plays a crucial epistemic role, the San Francisco Bay model and the Phillips–Newlyn machine. The San Francisco Bay model is a huge 1.5 acre model simulating the tidal and river action in the San Francisco Bay. The Phillips–Newlyn model is a 3-D hydraulic model of a macroeconomy in which colored water flows and accumulates in a system of tanks and channels. In the case of these two models, the philosophical intuition seems not to find any problem in locating the model system: it is the concrete thing! For example, Weisberg (2013) proposes, among many others, that a mathematical model is an abstract structure, while a scale model is a concrete physical object. It then follows that in the case of mathematical models, the written equations, for instance, are so-called model descriptions, whereas in the case of the San Francisco Bay model, the technical drawings and pictures of it would qualify as model descriptions. From the perspective of the artifactual account, this introduces an asymmetry in the way we approach mathematical models and so-called concrete models. If we consider the representational modes and media with which different models are constructed, then the written equations should be on par with various materials and artifacts that are used in *building* the 3-D scale model. The various technical drawings of the 3-D model are in turn more like a commentary accompanying a mathematical model in which the modeling decisions and assumptions made are discussed.

What then is the abstract — or nonconcrete — dimension of these 3-D physical models? Such abstract dimension is easier to pinpoint in the case of the Phillips–Newlyn model, which embodies and renders visible economic ideas such as the principle of effective demand and the conceptualization of the economy in terms of stocks and flows (Morgan and Boumans 2004). As for the San Francisco Bay model, a lot of theoretical knowledge of various kinds was needed in its construction, yet it can more readily be seen as an imitation of the behavior of one particular target system. The model actually functioned as a concrete demonstration of the disastrous effects of the Reber dam plan that, as a result, was not undertaken (see Weisberg 2013).

(iv) Biological Model Systems

Biological sciences use a host of models whose medium is biological, such as model organisms and laboratory populations. Among the newest inhabitants in this group are synthetic genetic networks that are engineered from genes and proteins. They are typically built by using mathematical models as kinds of blueprints. First, a mathematical model is constructed in order to study some hypothetical mechanism and its properties, and then, second, a gene regulatory network is engineered on the basis of the mathematical model, and implemented in host cells, frequently the *E. coli* bacteria. Knuuttila and Loettgers (e.g., Knuuttila and Loettgers 2013) have studied such synthetic genetic circuits from a philosophical perspective. The first and most famous of them, the repressilator, consists of a negative feedback loop of three genes that repress each others' expression. While the San Francisco Bay model partly shares the same material with its target system (e.g., water and its salinity), the repressilator is of the same materiality as naturally evolved genetic networks, functioning moreover under the same constraints as any other genetic circuits within the cell. The same materiality was crucial for its epistemic functioning: it was constructed to study whether some circuit designs from engineering, already extensively studied by mathematical modeling, could be realized also by biological organisms. Hence, even though the material medium of the repressilator was biological, it simultaneously embodied an abstract theoretical idea, the negative feedback loop. Here, the conceptual theoretical idea merged with the material medium as the feedback loop was biologically implemented. It should also be pointed out that one goal of synthetic biology is to standardize biological parts such that they could be combined in a multitude of ways to fulfill new functions (e.g. Endy 2005). This research program entails an establishment of a kind of biological language for which purpose synthetic biologists have been developing a Synthetic Biology Open Language (SBOL). SBOL is composed of genetic vocabulary terms and graphics for the representation of possible biological designs.[5]

4.2 Constrained Constitution and Justification

The representational devices used in model construction provide scientist an access to the worldly systems, but for this access to be epistemically rewarding, the model has to be built in a particular way. The other side of

[5] https://sbolstandard.org. Accessed 03.07.2020.

the epistemic access that models provide is due to their constrained constitution. In order to give us knowledge, models need to be constructed in such a way that they enable scientists to study the questions they are interested in. Instead of conceiving models as representations, the artifactual account views them as *erotetic devices.* That is, they are approached as artificial systems of dependencies that are constrained in view of answering a pending scientific question, motivated by theoretical and/or empirical considerations. What this means is that models are always already embedded in our existing world knowledge, and not separate entities in need of connection to worldly systems by a relation of representation. Consequently, the epistemological puzzle of how the representational relationship between a model and the world should be analyzed becomes that of studying how the model construction facilitates the study of some pending scientific questions.

The idea of constraint is central for the philosophical discussion of idealization. It has addressed the constrained nature of models in two different ways. On the one hand, it has paid attention to how models are designed to isolate some relevant or difference making features of the target system by disregarding and/or distorting the rest (cf. Strevens 2008; Cartwright 1999; Mäki 2011). On the other hand, it has been recognized that some idealizations are needed for tractability reasons and are entailed by the mathematical and statistical methods used (cf. McMullin 1985; Rice 2018). In other words, the discussion of idealization has highlighted both the enabling and limiting aspects of the constrained character of models. Part of this simultaneously enabling and limiting nature of models can be directly attributed to the affordances and restraints of the representational tools used. Furthermore, representational tools used go often hand in hand with the questions asked, or answers sought. Modeling is typically both method-driven and outcome-oriented (cf. Knuuttila and Loettgers 2017).

The Lotka-Volterra model provides a good example of the erotetic character of models, and its interplay with available representational tools. Volterra's version of the model is often traced back to the special characteristics of post-World War I fish populations in the Adriatic Sea, as if Volterra primarily sought to theoretically explain that particular phenomenon (e.g., Weisberg 2007). Knuuttila and Loettgers (2017) show that the model was actually a result of a longer-term research program. Already more than decades before the publication of the model, Volterra was interested in introducing the methods of physics to biology and economics, in particular the differential calculus, and what he called "the method of hypothesis." It does seem that the fluctuations in the predator and prey populations gave him a good case to explore such oscillations by using mathematical tools. The question he was interested in studying was whether the interdependence of the two populations was able to produce oscillations in their sizes. The ecologists of his time often attributed such fluctuations to environmental changes. To study the dynamics between the predator-prey populations, Volterra constructed a hypothetical system consisting solely of 'the intrinsic phenomena due to the voracity and fertility of the co-existing species' (Volterra 1927, 68).

One can approach Volterra's version of the Lotka-Volterra model as a purposefully designed, and patently artificial, epistemic tool. Volterra was not even remotely interested in the realistic depiction of any particular predator and prey system. Instead, he distinguished theoretically between 'external' and 'internal' causes, and wanted to study the interplay of some causes that he considered as 'internal'. Those causes would have "[…] periods of their own which add their action to these external causes and would exist even if these were withdrawn" (Volterra 1928, 5). Volterra's version of the Lotka-Volterra model was intimately connected to representational tools and the modelling methods he used. While some of the assumptions of the model were due to the way the model was constrained in order to answer the question concerning the fluctuations in the numbers of the predator and prey populations, others were due to the application of differential calculus to the problem of predation. Consequently, the assumptions that species increase or decrease in a continuous way makes them describable by using differential equations. Moreover, in order to make the model tractable, Volterra assumed that the individuals of each species are homogeneous and the birth and death rates are proportional to the number of living individuals of the species.

The case of the Lotka-Volterra model shows how the strategy of formulating a simple hypothetical system to study a well-defined question goes hand in hand with the application of particular mathematical tools and methods. Indeed, one conspicuous feature of modeling that has not been addressed by the representational approach is its use of transdisciplinary computational templates or methods (cf. Humphreys 2004; Knuuttila and Loettgers 2016). A computational template is a tractable formal device, like the Lotka-Volterra equations, that can be transferred across the disciplines to model various kinds of phenomena, exhibiting similar kinds of empirically observed patterns. Computational templates may have their origin in the formal sciences, like the network models in mathematics, or in specific research fields such as the Ising-model in the study of ferromagnetism, or the Lotka-Volterra model in population biology. Focusing on the prominent role of cross-disciplinary formal templates and methods in contemporary modelling practice requires a shift of perspective towards the artifactual dimension of scientific work. The representational approaches, the fictional approach included, approach models as targeting specific wordly phenomena, while the artifactual approach pays heed to the representational tools and methods that facilitate the study of various kinds of empirical and theoretical questions.

Finally, the critical question for the artifactual approach is that of how the various inferences, new results, and learning from models, more generally, can be justified. The allure of the representational approach to models is due to its seemingly clear-cut answer to the problem of justification: the model gives us justified knowledge if it resembles the target system more or less accurately, in relevant ways. Although neat, this solution is probably all too cheap: how is the model supposed to be compared to the worldly target systems, independently of our ability to represent them? In actual scientific practices the justification of models is two-fold. The justification is already partly built-in due to the already established resources — theoretical, empirical, mathematical, computational, and representational — utilized in model construction (cf. Boumans 1999). On the other hand the justification of a model is a result of the triangulation of different epistemic means: other models, experiments, observations and background theories. Such processes of triangulation are distributed in terms of epistemic labor, likely very complex and indirect, and usually inconclusive in character.

5 Conclusion

One of the paradoxes of the philosophical discussion of models and representation has been that it has largely overlooked the actual representational tools that scientists employ in model construction. In order to address the many languages of science — mathematical, diagrammatic, 3D, synthetic, and natural and written language — I have examined two recent approaches to models and representation, the DEKI-account of representation and the artifactual account. The DEKI-account recognizes different kinds of models, and groups them into concrete and nonconcrete models. I have argued that its treatment of nonconcrete models is in part misguided in that it separates the model system from its description. The model system is the imagined-object that is considered as the representational vehicle. The separation of the model system from its description leads to several problems concerning how the imaginings of scientists are generated and coordinated, how the enablings of various representational tools can be recognized, and how imagined-objects are supposed to represent. The artifactual account does not begin from the attempt of solving the problem of representation, but focuses instead on the representational tools used in modeling. These symbolic, semiotic, and material devices do not just provide an access to the problems scientists are interested in, but also draw together theoretical considerations, conceptual resources and empirical aspects of scientific work.

Funding Information This project has received funding from the European Research Council (ERC) under the European Union's Horizon 2020 research and innovation programme (grant agreement No 818772) and the Academy of Finland (grant No 290079).

References

Boumans, M. (1999). Built-in justification. In M. S. Morgan & M. Morrison (Eds.), *Models as mediators. Perspectives on natural and social science* (pp. 66–96). Cambridge: Cambridge University Press.

Bueno, O., & Colyvan, M. (2011). An inferential conception of the application of mathematics. *Nous, 45*, 345–374.

Bueno, O., & French, S. (2011). How theories represent. *The British Journal for the Philosophy of Science, 62*, 857–894.

Cartwright, N. (1999). *The vanity of rigour in economics: Theoretical models and Galilean experiments*. Centre for Philosophy of Natural and Social Science. Discussion paper series 43/99.

Chakravartty, A. (2010). Informational versus functional theories of scientific representation. *Synthese, 172*, 197–213.

Clark, A. (2008). *Supersizing the mind: Embodiment, action, and cognitive extension*. Oxford and New York: Oxford University Press.

Elgin, C. Z. (2004). True enough. *Philosophical Issues, 14*, 113–131.

Elgin, C. Z. (2010). Telling instances. In R. Frigg & M. Hunter (Eds.), *Beyond mimesis and convention: Representation in art and science* (pp. 1–18). Dordrecht: Springer.

Endy, D. (2005). Foundations for engineering biology. *Nature, 438*, 449–453.

French, S., & Ladyman, J. (1999). Reinflating the semantic approach. *International Studies in the Philosophy of Science, 13*(2), 103–121.

Frigg, R. (2010). Models and fiction. *Synthese, 172*(2), 251–268.

Frigg, R., & Nguyen, J. (2016). The fiction view of models reloaded. *The Monist, 99*(3), 225–242.

Frigg, R., & Nguyen, J. (2017). Models and representation. In M. Magnani & T. Bertolotti (Eds.), *Springer handbook of model-based science* (pp. 73–126). Berlin and Heidelberg: Springer.

Frigg, R., & Nguyen, J. (2018). The turn of the valve: Representing with material models. *European Journal for Philosophy of Science, 8*(2), 205–224.

Gelfert, A. (2016). *How to do science with models: A philosophical primer*. Cham: Springer.

Giere, R. N. (1988). *Explaining science: A cognitive approach*. Chicago and London: The University of Chicago Press.

Godfrey-Smith, P. (2006). The strategy of model-based science. *Biology and Philosophy, 21*(5), 725–740.

Godfrey-Smith, P. (2009). Models and fictions in science. *Philosophical Studies, 143*(1), 101–116.

Goodman, N. (1976). *Languages of art*. Indianapolis, IN and Cambridge, MA: Hackett.

Humphreys, P. (2004). *Extending ourselves. Computational science, empiricism and scientific method*. Oxford: Oxford University Press.

Humphreys, P. (2009). The philosophical novelty of computer simulation methods. *Synthese, 169*(3), 615–626.

Hutchins, E. (1995). *Cognition in the wild.* Cambridge, MA: The MIT Press.

Knuuttila, T. (2005). Models, representation, and mediation. *Philosophy of Science, 72*(5), 1260–1271.

Knuuttila, T. (2011). Modelling and representing: An artefactual approach to model-based representation. *Studies in History and Philosophy of Science Part A, 42*(2), 262–271.

Knuuttila, T. (2017). Imagination extended and embedded: Artifactual and fictional accounts of models. *Synthese.* https://doi.org/10.1007/s11229-017-1545-2.

Knuuttila, T., & Loettgers, A. (2013). Synthetic modeling and the mechanistic account: Material recombination and beyond. *Philosophy of Science, 80*, 874–885.

Knuuttila, T., & Loettgers, A. (2016). Model templates within and between disciplines: From magnets to gases — And socio-economic systems. *European Journal for Philosophy of Science, 6*(3), 377–400.

Knuuttila, T., & Loettgers, A. (2017). Modelling as indirect representation? The Lotka-Volterra model revisited. *The British Journal for the Philosophy of Science, 68*(4), 1007–1036.

Kress, G., & van Leeuwen, T. (2001). *Multimodal discourse: The modes and media of contemporary communication.* London: Arnold.

Levy, A. (2015). Modeling without models. *Philosophical Studies, 172*(3), 781–798.

Mäki, U. (2011). Models and the locus of their truth. *Synthese, 180*, 47–63.

McMullin, E. (1985). Galilean idealization. *Studies in History and Philosophy of Science Part A, 16*, 247–273.

Morgan, M. S., & Boumans, M. J. (2004). Secrets hidden by two-dimensionality: The economy as a hydraulic machine. In S. de Chadarevian & N. Hopwood (Eds.), *Model: The third dimension of science* (pp. 369–401). Stanford, CA: Standford University Press.

Morgan, M. S., & Morrison, M. (Eds.). (1999). *Models as mediators. Perspectives on natural and social science.* Cambridge: Cambridge University Press.

Morrison, M., & Morgan, M. S. (1999). Models as mediating instruments. In M. S. Morgan & M. Morrison (Eds.), *Models as mediators. Perspectives on natural and social science* (pp. 10–37). Cambridge: Cambridge University Press.

Parker, W. (2009). Does matter really matter? Computer simulations, experiments and materiality. *Synthese, 169*, 483–496.

Pincock, C. (2004). A new perspective on the problem of applying mathematics. *Philosophia Mathematica, 12*, 135–161.
Rice, C. (2018). Idealized models, holistic distortions, and universality. *Synthese, 195*(6), 2795–2819.
Salis, F. (2016). The nature of model-world comparisons. *The Monist, 99*(3), 243–259.
Salis, F. (2019). New fiction view of models. *British Journal for Philosophy of Science*. https://doi.org/10.1093/bjps/axz015.
Strevens, M. (2008). *Depth: An account of scientific explanation*. Cambridge, MA: Harvard University Press.
Suárez, M. (2003). Scientific representation: Against similarity and isomorphism. *International Studies in the Philosophy of Science, 17*, 225–244.
Suárez, M. (2004). An inferential conception of scientific representation. *Philosophy of Science, 71*, 767–779.
Toon, A. (2010). The ontology of theoretical modelling: Models as make-believe. *Synthese, 172*(2), 301–315.
Toon, A. (2012). *Models as make-believe: Imagination, fiction and scientific representation*. Chippenham and Eastbourne: Palgrave Macmillan.
Volterra, V. (1927). Variations and fluctuations in the numbers of coexisting animal species. In F. M. Scudo, & J. R. Ziegler (Eds.), 1978, *The Golden Age of Theoretical Ecology: 1923–1940* (pp. 65–236). Berlin: Springer-Verlag.
Volterra, V. (1928). Variations and fluctuations of the mumber of individuals in animal species living together. *Journal du Conseil International pour l'Exploration de la Mer, 3*, 3–51.
Vorms, M. (2011). Representing with imaginary models: Formats matter. *Studies in History and Philosophy of Science, 42*, 287–295.
Walton, K. (1990). *Mimesis as make-believe: On the foundations of the representational arts*. Cambridge, MA: Harvard University Press.
Weisberg, M. (2007). Who is a modeler? *The British Journal for the Philosophy of Science, 58*(2), 207–233.
Weisberg, M. (2013). *Simulation and similarity: Using models to understand the world*. New York: Oxford University Press.
Weisberg, M., & Reisman, K. (2008). The robust Volterra principle. *Philosophy of Science, 75*(1), 106–131.
Zhang, J. (1997). The nature of external representations in problem solving. *Cognitive Science, 21*, 179–217.

Part IV

Language in Mathematics and in Empirical Sciences

8

On Mathematical Language: Characteristics, Semiosis and Indispensability

Jesus Alcolea

1 Focus on the Problems

Mathematicians and other people often discuss mathematics as a language, a universal language. At the same time, it is said that mathematics holds a special status among the sciences. In particular, (the language of) mathematics is the language of science. In some way, mathematics is the basis of the physical world, but globally it is beyond any other science. In other words, it is not a mere servant of the sciences.

Apparently, mathematical language is simple, with a little grammar and a limited vocabulary, but very different from other languages. Unlike natural languages, it is a rigorously defined and unambiguous language. This characteristic constitutes its greatest advantage: its complete lack of ambiguity. As a formal language, it cannot express a wide variety of

J. Alcolea (✉)
Department of Philosophy, Faculty of Philosophy, University of Valencia, Valencia, Spain
e-mail: jesus.alcolea@uv.es

W. J. Gonzalez (ed.), *Language and Scientific Research*,
https://doi.org/10.1007/978-3-030-60537-7_8

things, but can make precise and unambiguous statements about vaguely determined things.

Although the language of mathematics is limited in the range of things that can be expressed in it directly, it can be adapted to the needs of a particular application extending its usefulness greatly and posing the main challenge in its application. This challenge is primarily linguistic, not mathematical, in nature.

One of the limitations in this language is the fact that the notion of time is absent from it completely. Time and dynamic processes can easily be and often are modeled mathematically, but this is part of the adaptation of the language of mathematics to the particular application in question.

A *language* is a medium for expressing or communicating verbally or visually facts, opinions, thoughts, feelings, desires, commands, etc. (cf. Baber 2011). Each language employs abstract symbols (verbal or visual) to represent things. Mathematics exhibits these characteristics as a language, albeit with different emphasis and importance. The range and distribution of topics communicated in natural languages and those communicated in mathematical language differ in some significant ways. Feelings and emotions are not expressed in mathematical terms, and imprecisely defined terms are not allowed in mathematical language.

The formal character of mathematical language facilitates analysis and reasoning through the mechanical manipulation of symbols, according to precise rules. In this way, what mathematicians do is reasoning on abstract objects, a task characterized as logical, and this reasoning is frequently included into the modeling of objects, facts or events we find in the physical world that is studied by different sciences and technics. With the resulting models, scientists make descriptions and predictions with the aim of understanding the world we live on.

The aim of our contribution will be to analyze some of the characteristics of mathematical language, the role of semiotic modes in reproducing the effectiveness of mathematics in science, and the relation of the problem of indispensability to the one of the reasonable effectiveness.

2 Characteristics of Mathematical Language

It is obvious that one's understanding of language is greatly enhanced by a knowledge of its main grammatical elements. This is particularly true of mathematical language due to the complicated structure of many of the sentences of advanced mathematics, and to the fact that these sentences are supposed to be precise and without the ambiguities proper of the natural language.

The mathematical complexity contrasts with the relatively small number of basic concepts and terms we commonly use to express all mathematical sentences. But it does not mean that mathematical texts are written in a fully formal symbolism, which would be difficult to read. It is customary to write them using also the colloquial language, avoiding impreciseness. The ideal is to write in the most approachable way making sure that the possible reader will be able to follow the lines and make them more formal if circumstances so require. And here is where logic comes in: when a mathematical argument or proof is difficult to grasp the best and most convincing way of showing that is valid or correct is to present it formally. The process can be reversed. It would be extremely easy to eliminate the formalism, replacing every symbol with words and phrases of the natural language, and the result would still be a proof. In fact, it is the *same* proof. The chosen language, whether symbolic, verbal, or even visual, might affect the length of the proof, or the ease with which one can understand it, but it does not affect whether the argument does or does not constitute a proof. In human terms, as K. Devlin (1998, 51–52) says, "being a proof means having the capacity to completely convince any sufficiently educated, intelligent, rational person, and surely that capacity has to do with some kind of abstract pattern or abstract structure associated with the argument."

On the other hand, the complexity is generally associated to specialized domains of activity. These domains have specialized vocabularies and languages, which enable practitioners to communicate efficiently about the subjects studied in them. The problem is that it may exclude people who have not been introduced to this type of practice, as it is particularly the case with mathematics. Its high degree of abstraction may be an obstacle not only to communicate with other people but even to produce and transmit new mathematical knowledge. So, it is really important to be very precise about what mathematical language is. The identification with formal systems is the easy way to do it. It is enough to acknowledge a vocabulary or set of formal symbols and a set of rules that govern the construction of well-formed statements or formulas. For some, the mathematical activity and communication hinge in part on the formation and transformation of sequences of formulas. Others have also recognized other semiotic systems as for example visual resources such as graphs and diagrams that play an important role in the process of producing and communicating mathematical knowledge. Linguistic and sociolinguistic research has exerted a growing influence on all issues related to specialized languages. In particular, M.A.K. Halliday started studying the interactions between linguistics and mathematical education in 1974. His work had had considerable influence in the application and development of semiotical theories in mathematics, and recently in multimodal semiotics that centers on the role of multiple modes of communication as gestures and visual interactions.

But what do we find in the characteristics of mathematical language?

(1) The special *vocabulary* used to name mathematical objects and processes. It includes proper mathematical words as asymptote, logarithm, parallelepiped, trigonometry, hyperbola, etc., and common words with natural language as maximum, minimum, set, field, multiply, point, etc., originated in non-mathematical contexts, sometimes with a bit different meaning and adapted for rigorous mathematical purposes. Difficulties in mathematical teaching and learning are frequently related to the use of such words in the appropriately mathematical way.

(2) The development of dense *set of words* such as *minimum common multiple, tangent to a parameterized curve* or *logarithmic curve between two points.* In order to be understood the set must be taken together as a single unit of meaning. This unit will provide the condensed information and the required combination into longer mathematical sentences. The complexity to be achieved will make possible to handle higher mathematical concepts.

(3) The transformation of processes into objects. The linguistic way to do it is by forming a substantive out of a verb. For example, transformation out of transform, integration or integral out of integrate, as in Laplace transformation or Fourier transformation, and in numerical integration or symbolic integration, or Riemann integral, etc. Sometimes the verb remains substantivized, as in Laplace transform. This objectification of a process serves at least two purposes: to remind us our thinking about the nature of the mathematical activity, and the way in which people may relate to mathematics. The consequence is that the mathematical agents or practitioners become volatilized as if human agency never existed in mathematics, as if nobody were engaged in fruitful mathematical discourse. Therefore, it is apparent that mathematical discourse is a discourse without discursive subjects, and mathematics has nothing to do with rhetoric. On the other hand, and as an important aspect of the process of thinking in mathematical terms, this transformation of processes into objects allows the creation of new mathematical objects that incorporate the processes; and the ability to think about some mathematical ideas as, for example, the idea of function, both as a process and as an object. By themselves, these processes or objects can be subject to other (iterative) processes as, for example, multiplication, derivation, integration, differentiation, etc.

Anna Sfard (2008, 44) has explained that this process of objectification involves "two tightly related, but not inseparable discursive moves: *reification*, which consists in substituting talk about actions with talk about objects, and *alienation*, which consists in presenting phenomena in an impersonal way, as if they were occurring of themselves, without the participation of human beings." However, it does not mean that many of the characteristics of mathematical language have no important roles in enabling mathematical activity. In fact, according to Sfard, there is no distinction between communicating and thinking, because inspired by Lev Vygotsky and Ludwig Wittgenstein she defines thinking as a form of communication, and therefore thinking and doing mathematics remain identified in mathematical discourse when communicating mathematically with other people.

In focusing on features of verbal language, Candia Morgan (2014) reminds us the importance of the roles played by other semiotic systems in the development of mathematics. A prime example to consider is the way in which Descartes' algebraization of geometry transformed the field. Algebraic notation allows the manipulation according to formal rules that produce new forms of statements that provide new mathematical knowledge. In contrast, and by their very constitution, although graphical forms do not allow this sort of manipulation, they provide a more holistic and dynamic understanding of the objects represented.

Elements in mathematical texts which function as discernible units through systematic choices from the grammars of language, visual images and mathematical symbolism include graphs, diagrams, tables, stretches of linguistic text and the symbolic equations, as well as photographs, maps and other forms of drawings. O'Halloran (2005) has analyzed the representation of different modes of communication — verbal, algebraic, tabular, diagrammatic, and graphical — and the conclusions are that these modes will deliver different sorts of information on the "same" mathematical object. One can well imagine oneself reflecting, for example, on the aspects and on the actions one can perform when faced with a function expressed in verbal, algebraic, tabular, diagrammatic or graphical form.

Morgan (2014) says that the differences between possibilities presented by different modes may contribute the development of mathematical knowledge through *conversion*. That is, converting from one mode to another implies "understanding and coordinating the mathematical structures of both modes" (p. 390). For example, laying out the graph of a function given in algebraic form or formulating the algebraic equation for a given graph. New technologies applied to mathematics have allowed to design environments in which the different semiotic forms of representation are epistemologically relevant.

The explanation offered so far of the characteristics of mathematical language should not make us think that there is only one variety of this language. It is quite obvious that children studying mathematics cannot use the specialized language as scholars do. In addition, Burton and

Morgan (2000) have noted variations in the linguistic characteristics of various mathematical publications. Some researchers take into account not only strictly mathematical contexts and fields, or their own status as practitioners, but also educational contexts and fields. On the other hand, it is also obvious that the different purposes pursued with mathematical communication require the use of different forms of language. Thus, for example, to count the steps followed to solve a mathematical problem we will use a language whose characteristics will be different from that used to present the rigorous proof of a conjecture. Therefore, the best we can do is consider mathematical language as a set of forms of language sharing the characteristics we have been talking about, and the specialized features to identify them as mathematical.

3 The Perspective of the Multimodal Social Semiotics: Language, Symbolism and Images

Multimodality explores how the individual and combined meaning-making potential of different semiotic resources defines and is defined through communication. Jewitt understands multimodality as "a field of application" that invites "a variety of disciplines and theoretical perspectives" (Jewitt 2009, 2). Some of the key foci in multimodality research are represented by the semiotic potential and use of individual modes, the intersemiotic relations, the connections between multimodality and technology, or the connections between knowledge, pedagogy and literacy. O'Halloran (2009, 113) demonstrates that the "historical view of the semiotic landscape in mathematics reveals how mathematical knowledge evolves and materialises as multimodal and semiotic objects, activities and events constructed using semiotic resources which are shaped by social and scientific technologies."

Here mathematics will be considered as a multimodal semiotic process integrating language, images, and mathematical symbolism to create knowledge. The integration of the linguistic, visual, and symbolic systems leads to the semantic development of mathematics beyond the limits of other forms of communication. It is obvious that this development has resulted in new domains of knowledge, and has had an enormous success in the natural sciences. That's why it is not difficult to think in conceptualized mathematics as a multimodal semiotic process designed to go beyond the everyday experience to reach an abstract semiotic domain where to restructure thought and reality in order to explain how mathematics achieves its functional character, and becomes indispensable and effective.

According to O'Halloran (2015), there are three main ideas to be taken into account in this perspective: (1) Sign systems act as tools to structure thought and reality. (2) The development of new integrated, written systems which combined textual forms — linguistic, symbolic — with visual forms — graphs, diagrams or figures — allows to construct new views of the world. (3) As a multimodal hierarchical epistemic system, while it is useful for describing and predicting events in the natural world, mathematics has some limits to model and predict the socio-cultural world. These ideas present a major challenge insofar as the aim is to develop new semiotic tools for modeling the social world "to replicate the *effectiveness* of mathematics in the natural sciences" (2015, 288; italics is ours).

3.1 Creating a Scientific View Through (Natural) Language

As Halliday (2004, xvii) writes, "the grammar of every language contains a theory of human experience; it categorizes the element of our experience into basic phenomenal types, construing these into configurations of various kinds, and these configurations in turn into logical sequences." Illustrating and discussing the difficulties that are characteristic of scientific English, Halliday shows that natural language creates a scientific view of the world. Drawing on various academic sources, writing in primary schools, high-school textbooks, studies on metaphor, mathematical or scientific texts, papers or lectures, or works by Chaucer, Newton, Priestley, Dalton, Darwin and Maxwell, he points to a major semantic shift from a commonsense view of the world to an abstract scientific view. As examples for us, he finds useful these *categories*:

(1) Some mathematical terms as circle, center, radius, diameter or circumference are presented through *interlocking definitions*: "The diameter of the circle is twice the radius."

(2) In the natural sciences, technical concepts derive their meaning from being organized into taxonomies. These are highly ordered constructions in which every term has a definite functional value. For instance, a *technical taxonomy* is typically based on two fundamental semantic relationships: '*a* is a kind of *x*' (superordination) and '*b* is a part of *y*' (composition).

(3) Some expressions used in mathematical language are *special expressions*, because they have a *special (technical) grammar*. For instance, "solving the open sentence over *D*." This is the (special) expression to be defined, instead of each one of its words: "If *D* is the domain of a variable in an open sentence, the process of finding the truth set is called *solving the open sentence over D*." Halliday says that this kind of special grammar is more common in mathematics than in science, and "mathematicians have often had to stretch the grammar a little in order to say what they want" (2004, 167).

(4) The *lexical density* is a measure of the density of information identified as the number of lexical words per clause, and according to how tightly the key items appear in the grammatical structure. For instance, "A *parallelogram* is a *four-sided figure* with its *opposite sides parallel*."

(5) *Syntactic ambiguity* is practically impossible to find in mathematical language, but this is not the case in scientific writing which may focus attention on some verbal expressions as "may be reflected (in)," "are … associated (with)," "means," that are ambiguous in two respects: we cannot tell whether they indicate a relationship of cause or of evidence, and supposing that we can identify a relationship of cause, we still cannot tell which causes which. In the sentence, "Lung cancer death rates *are* clearly *associated with* increased smoking," the expression in italics can face in either direction: either 'cause' or 'are caused by.' In addition, what does *lung cancer death rates* mean? Is it 'how

many people die from lung cancer,' or 'how quickly people die when they get lung cancer'? Or is it perhaps 'how quickly people's lungs die from cancer'? (2004, 170).

(6) The metaphor is a great driving force to the overall power of language, and particularly the *grammatical metaphor* has the most significant effects due to its applications and for its potential to increase the power for *theorizing*: "Grammatical metaphor — says Halliday (2004, xvii) — creates virtual phenomena — virtual entities, virtual processes — which exist solely on the semiotic plane; this makes them extremely powerful abstract tools for thinking with." Grammatical metaphor is found particularly in scientific discourse from ancient times, and although it has spread across many different registers of language, in English at least the main impetus came from the languages of science. After Newton, scientific writers favored this metaphorical mode of expression writing, for instance, 'this event caused that event' instead of 'this happened, so that happened.' Halliday sees the reason for this in the nature of *scientific discourse*. Newton and his successors created a new English at the service of a new kind of knowledge, in which experiments were made, and some general principles were derived with the help of reasoning and mathematics, principles to be tested later by means of other experiments. From here, says Halliday (2004, 174),

> [t]he discourse had to proceed step by step, with a constant movement from 'this is what we have established so far' to 'this is what follows from it next;' and each of these two parts, both the 'taken for granted' part and the new information, had to be presented in a way that would make its status in the argument clear. The most effective way to do this, in English grammar, is to construct the whole step as a single clause, with the two parts turned into nouns, one at the beginning and one at the end, and a verb in between saying *how* the second follows from the first.

The use of (scientific) language embodies theory, a theory that becomes more abstract and more powerful, and the product of a conscious design. And through grammatical metaphors the commonsense picture of the world is transformed into one that imposed regularities and control on the experience. The new vocabulary of technical terms in scientific language produces the most profound grammatical changes, and with them the virtual entities logically interrelated transform human view of the world. As mathematics was taken as the base of science and the instrument for justifying its findings and deriving its results, mathematical language and symbolism assume the task of restructuring the world.

(7) *Semantic discontinuity* points to semantic leaps that writers make and that readers are expected to follow to reach a determined conclusion. It is not uncommon to find this sort of discontinuity in scientific writing, and the reader has to discover new information and get it ready for understanding something else that depends on it, or use it as a point of departure for the next step in an argument.

To sum up, the everyday experience as a series of events was replaced with a virtual world of cause and effect. Linguistically, as O'Halloran says (2015, 291), "the abstract scientific view was constructed by recoding processes (i.e., verbs) as metaphorical (virtual) entities (i.e., nouns), and logical connections (i.e., conjunctions) as causal processes (i.e., verbs), in order to relate the virtual entities to each other."

3.2 From Rhetorical Forms to Actual Symbolic Systems

It seems that mathematical symbolism evolved from linguistic rhetorical forms to abbreviated forms which contained linguistic elements and symbols, to the actual symbolic systems. It seems also that the rhetorical character of many sixteenth-century texts was in fact a conscious choice of how to organize the new material in terms of an older tradition. It points to the idea that rhetoric might have played an important role in the very creation of modern mathematics, and that the rhetorical character of the new genre of mathematical texts should be seen as a paradigm of the 'art of thinking' (cf. Cifoletti 2006).

However, modern mathematical notation developed in written form, and new visual-based strategies were developed. In fact, as O'Halloran comments (2015, 293), "the printing press, and the subsequent shift to standardized forms of symbolic notation, paved the way for the development of mathematical symbolic notation as a semiotic tool developed for written (and today digital) modes of communication." The mathematical symbolism was transformed for different purposes and, in particular, to serve as a basis of arguments on scientific findings, contributing both to the transformation of scientific vision of the world and to the grammatical system for encoding meaning. In consequence, the scientific writing created multiple configurations of virtual entities and processes that interacted in different ways and that reinforced (logical) thinking and reasoning. In other words, as O'Halloran keeps saying, "mathematics organizes experience in terms of the relations between multiple (generalized) entities interacting with each other via multiple (generalized) processes at any given instance." The derivation of results was possible rearranging in an easy way the configurations that were encoded in an economical and unambiguous fashion using positional and spatial notation, specialized symbols, operations, brackets, etc. These resources created a dynamic representation of multiple relations easy to be comprehended and manipulated, and mathematical symbolism extended its meaning potential for capturing patterns and relations. In this way, mathematics was "designed to manage the complexity of the universe, as the ultimate tool for mapping and thinking about patterns and relations," and its language and symbolism imposed "a certain way of experiencing reality" (2015, 294). As both of them were related to images, it was created a semantic connection between mathematical language, symbolism and image as a first step towards multimodal semiosis.

3.3 The Function of Images to Create Reality

Images and writing have been combined through the ages. O'Halloran (2015) quotes the study of paintings by Berger (1972) to show how images, as natural language and other semiotic resources, function to structure thought and reality, and to create the world around us, rather than reflect it. The consequence is that, as other scholars (cf. Bateman 2014) have showed, images provide experiential, logical, interpersonal, and textual meanings, all of which are related as a series of parts to the whole, and to the way people perceive world. "In particular — O'Halloran (2015, 295) says — mathematical images (e.g., graphs, diagrams, and other visualizations) are powerful resources for reasoning because they permit mathematical entities, processes, and circumstances to be viewed in relation to each other, offering a powerful lens to understand patterns and relations," and for deriving mathematical results (cf. Davis 1974).

O'Halloran explains also how mathematical discourse evolved from language. Early algebra was *rhetorical* algebra with linguistic descriptions and solutions. But that discourse also evolved from geometrical images, which were less abstract than the contemporary ones. For instance, Descartes includes not just a series of line segments and curves describing the path of a ball through water, but also a seventeenth century gentleman with a sort of racket, standing at the edge of water and looking at the ball (2005, 43) (Fig. 8.1)[1]:

Then O'Halloran explains how Descartes mistrusts language, as he feels it belongs to the commonsense world of perception, and therefore develops the algebraic symbolism to try and reduce the natural language

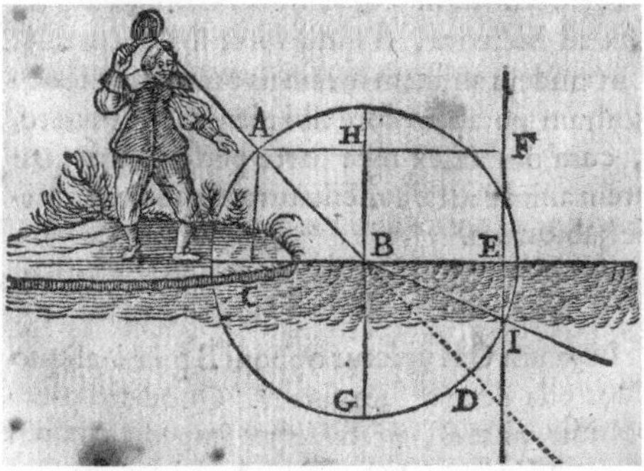

Fig. 8.1 Descartes on geometrical images with the setting

[1] This image comes from Descartes (1682) p. 226. The author is grateful to the Biblioteca Històrica (Historical Library), Universitat de València, for the right to reproduce it.

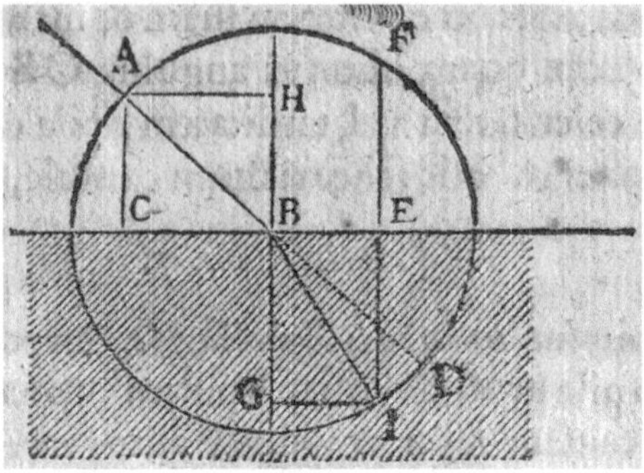

Fig. 8.2 Descartes on geometrical images without the setting

to the minimum. In this way the images in geometrical drawings are gradually reduced, first by removing the setting and showing only the hand or the eye of the observer, then suppressing the observer from the picture (2005, 44) (Fig. 8.2)[2]:

The evolution of the semiotics reveals not only how the human beings were gradually removed from mathematical images but even the situational contexts, in a way that everything turns around the mathematical objects themselves. Descartes' union of algebra and geometry to become analytic geometry supposed an important shift from the natural world to the abstract world so clearly observable in Newton's mathematical writings. In this way, mathematics was transformed in a semiotic resource that combines different sign systems, showing that the "changes of languages that have the form of *visualization* or *symbolization* are closely related" (Kvasz 2008, 89). As it is actually used, mathematics assigns different roles to language, symbolism and images. Language serves for introducing, contextualizing and describing problems, and symbolism is used for working towards the solution of problems by means of reasoning. Images play an important role, "because the parts are seen in relation to the entire mathematical construct, opening up new avenues of reasoning. Significantly, mathematical images also relate to our lived-in sensory experience, providing a bridge from perceptual understanding of the

[2] This image comes from Descartes (1682) p. 228. The author is grateful to the Biblioteca Històrica (Historical Library), Universitat de València, for the right to reproduce it.

world to the abstract semiotic realm of mathematics, which in turn lead to further abstractions, given the close links between the images and the mathematical notation" (O'Halloran 2015, 296).

Anyway, the three semiotic modes combine in actual mathematical discourse, and work simultaneously on three levels: *ideationally*, by representing in some way what is going on in the world, *interpersonally*, by creating interaction between communicating parties, and *textually*, by weaving ideational and interpersonal meanings into a textual whole. The ideational function subdivides into two functions: one *experiential* and one *logical*. The first construes experiences, and the second construes logical relations between experiential meanings. Moreover, the interpersonal functions (e.g. informing, questioning, and direct or indirect address) can be realized by the grammatical systems of mood (indicative, interrogative, imperative) and person (the 'indirect' third person, and the 'direct' first and second person). So, for O'Halloran, mathematical discourse is not just a resource for abstractly representing and performing logical operations on aspects of the world, but also as a way of addressing people, through its use of a *rhetoric of commands* (2005, 104) ('let's suppose that,' 'let us consider that,' 'calculate,' 'find,' etc.), and of objective, impersonalized images:

> Indeed, it is difficult to imagine how big problems in the world today can be resolved, unless the focus on the experiential and logical domain is expanded to include the interpersonal domain, and this necessarily involves bringing together science, mathematics, and the humanities, aided by digital technology. From there, it may be possible to understand human and physical worlds as an integrated system. (O'Halloran 2015, 300)

In fact, human subjectivity is not completely removed, becomes part of the language under the form of a point of view and is expressed in the language. (cf. Kvasz 2008, 128–131).

4 Indispensability and Effectiveness of Mathematics

Mathematics can be applied to the world through some descriptions of the world, using the language as a medium in a way that has taken us to consider the role of mathematics as language, and in a way that makes those uses of language as "unreasonable effective." So the view of mathematics as language allows us to understand the mechanisms for the applicability to be effective (cf. Sarukkai 2005).

As already emphasized, one of the most important features of mathematics is its applicability to virtually all branches of empirical science in a variety of ways and allowing elegant and economical propositions. So important is mathematical language to science that it is practically impossible to imagine how scientific theories could be formulated without it. The idea that the 'Book of Nature' is written in mathematical language has been credited through the history of Western science and been confirmed by the success of scientific theories. But the success is one thing, and the applicability of mathematics is another. As mathematics suffered important transformations, practitioners were perplexed with the concordance between mathematics and science, and the perplexity was transformed into a sort of puzzle at the possibility of applying mathematics to the empirical world. This puzzle about the mathematical character of science has been related to problems on the indispensability of mathematics to certain kinds of scientific explanation and to reflections in the debates about mathematical (or scientific) realism.

So, from the uncontroversial fact that mathematics is indispensable to science, Quine (1980, 1981) and Putnam (1979) have argued that the indispensability provides good reasons to believe in and defend the existence of mathematical entities. Although in contemporary philosophy of mathematics there is different versions of the argument (cf. Panza and Sereni 2015), the so-called *Quine-Putnam indispensability argument* establishes that reference to mathematical entities — sets, numbers, functions, etc. — is indispensable to our best scientific theories, and so we must be committed to their existence. Moreover, from an epistemological point of view mathematical entities are on a par with the scientific entities, because our belief in them is justified by the confirmational evidence of the theory as a whole.

According to an indispensability argument, we must believe certain propositions because it is indispensable for achieving certain purposes (Field 1989, 14). For our present goal, in some way doing science, the scientific indispensability argument concerns with our belief in the existence of a class of entities, and this belief is indispensable to our best scientific theories. For instance, in his paper 'Success and limits of mathematization,' Quine (1981, 149–150) wrote:

> Ordinary interpreted scientific discourse is as irredeemably committed to abstract objects — to nations, species, numbers, functions, sets — as it is to apples and other bodies. All these things figure as values of the variables in our overall system of the world. The numbers and functions contribute just as genuinely to physical theory as do hypothetical particles.

According to it, abstract (mathematical) entities are as indispensable to our scientific theories as theoretical entities of our best physical theories. Notice however that, to be strict, the indispensability is about the *postulation of*, and the *reference to*, entities which are *posits*. In fact, "[p]hysical objects, small and large, are not the only posits," but "the abstract entities which are the substance of mathematics (…) are another posit in the same spirit." Due to his empiricism, for Quine, the total scheme of science — *mathematical, natural, human* — "must be kept squared with experience" (1980, 45), in a way that the segments of mathematical science employed by empirical science enjoy the empirical support enjoyed by this science as a whole. This is the proverbial *confirmational holism*, the view that theories are confirmed or disconfirmed as a whole. The confirmation of a theory by empirical discoveries implies the confirmation of the whole theory, and so the confirmation of the mathematics used in the theory.

We now turn our attention to the important issue of the *unreasonable effectiveness of mathematics* to natural sciences. It is connected with the problem of the indispensability: although mathematical methods can be *a priori* and driven by aesthetic considerations, they are unavoidable in describing and explaining the physical world. The physicist Eugene Paul

Wigner once wrote "that the enormous usefulness of mathematics in the natural sciences is something bordering on the mysterious and that there is no rational explanation for it" and that "[t]he miracle of the appropriateness of the language of mathematics for the formulation of the laws of physics is a wonderful gift which we neither understand nor deserve" (1960, 2 and 14). Wigner knew that Kant had left written that "since in any doctrine of nature there is only as much proper science as there is a priori knowledge therein, a doctrine of nature will contain only as much proper science as there is mathematics capable of application there" (1786, 6). Steven Weinberg, the Nobel Prize-winning physicist, remarks also that physicists "generally find the ability of mathematicians to anticipate the mathematics needed in the theories of physics quite uncanny," and qualifies the applicability of mathematics quite amazing: "It is very strange that mathematicians are led by their sense of mathematical beauty to develop formal structures that physicists only later find useful, even where the mathematician had no such goal in mind" (1993, 157). But it is not unreasonable that mathematical concepts return the favor of the applicability of mathematics because the ultimate source of those concepts is experience (cf. Steiner 2005, 632).

According to Wigner, the considerations that mathematical concepts are interesting, ingenious or beautiful, and "turn up in entirely unexpected connections," shape mathematics in all its levels of complexity, and not only elementary mathematics which was "formulated to describe entities which are directly suggested by the actual world." However, mathematical concepts are not chosen for their simplicity or for their empirical adequacy, but "for their amenability to clever manipulations and to striking, brilliant arguments." That is why some parts of some mathematical theories are more attractive than others for some scientists in order to elaborate some scientific theories, although it does not mean that these theories could be the correct ones. In addition, when applying mathematics to scientific problems, Wigner says with Popperian flavor that, in the end, we often find that "we 'got something out' of the equations that we did not put in" (1960, 2, 7 and 9). This question explains the versatility and usefulness of mathematical language, and certain perspective on the realist philosophy of mathematics.

It seems that the *puzzle* — as Mark Colyvan (2009, 689) calls it — of the 'unreasonable effectiveness' of mathematics through its applications does not depend on any particular philosophy of mathematics, but on what one means by 'application' or 'applicability' in this context, and because mathematics "is applicable in many senses," according to Mark Steiner (1998, 2005, 1). When understood semantically, the problem of applicability may be reduced to the contextual interpretation of the vocabulary in mathematical propositions, and it was solved by Gottlob Frege: "never to ask for the meaning of a word in isolation, but only in the context of a proposition" (1953, xxii). However, this German thinker didn't explain the appropriateness of mathematical concepts for describing the world, or the active role that mathematics play in the discovery — and not only as a framework for the statement — of the correct scientific theory.

According to Steiner, the strategy that physicists have followed for discovering new theories depends on mathematical *analogies* between past theories and new proposals. The strategy is *Pythagorean* when it is expressible in the language of pure mathematics and following the characteristics of an object, and it is *formalist* if it depends on the syntax of mathematical language. Newton or Maxwell had already used these analogies. But in general scientists used them "because they had no real alternative, [and] looked for laws bearing a similar (not necessarily identical) mathematical form to the laws they were trying to augment, refine, or even replace" (1998, 3).

In this point, it is relevant to mention that, for Ivor Grattan-Guinness (2008), Wigner's appeals to the beauty of mathematics and to the manipulability of mathematical expressions cannot serve as a ground for mathematics nor an explanation of its development or importance. Moreover, it neglects "clear indications from history of sources of both reasonableness and effectiveness of the natural sciences in mathematics" (2008, 15). It could be testified by a great number of examples in George Pólya's *Mathematics and Plausible Reasoning* (1954) and *Mathematical Methods in Science* (1963) or Hermann Weyl's *Philosophy of Mathematics and Natural Science* (1949). For instance, Weyl explains the formation of concepts and theories in connection with mechanics. In fact, the most significant physical concepts are obtained in accordance with the scheme of mathematical abstraction. So, elaborating on Bacon's formula 'dissecare naturam,' and Descartes' idea that in order to reach certainty and evidence the mathematician starts with "the easiest and simplest," Weyl recognizes the following phases of dissection into simple elements: (1) The dissection of the three-dimensional spatial reality allows us "forming an intuitive spatially isolated and relatively constant unit" (1949, 145–146). (2) The spatio-temporal coincidence and amalgamation of several simple phenomena result in the conception of an intuitively experienced event. (3) The apperception of the 'being-so,' highlighting the characteristic features of the phenomena, provides "the grouping together of similar things," that is, the subordination or classification under concepts, and the formation of 'natural' classes. (4) Beyond isolable elements, it can be interpreted "a series of properties which always appear together as an indication of a concealed something," giving rise to hypothetical elements (forces, etc.). Simultaneously with the dissection, certain principles about the union of elements into a whole must be set (formation of the resultant of forces, etc.), and can be quantitatively expressed by mathematical functions into propositions (cf. 1949, 145–147). According to Weyl, this is "the constructive character of the natural sciences" and, with a distinctly Quinean flavor, "[i]ndividual scientific statements cannot be ascribed an intuitively verifiable meaning, but truth forms a system that can be tested only in its entirety" (1949, 151).

On the other hand, Grattan-Guinness also collects seven categories or ways in which a new scientific theory may relate to an old one: *reduction*, *emulation*, *corroboration*, *importation*, *revolution*, *innovation* and *convolution*. He defends that this categorization makes sense of the *analogies* that operate in many scientific disruptions. In fact, with an ever-increasing number of mathematical theories and "an impressive tableau of ubiquitous topics and notions, theory-building can be seen as reasonable to a large extent." However, the effectiveness may come in need of some improvement after further, if perceived, "desimplifications." So, "[i]nstead of 'effective but unreasonable,' read — continues Grattan-Guinness (2008, 15) — 'largely reasonable, but how effective?'" This motto can also guide evaluations of (un)reasonable (in)effectiveness in other contexts related as the notational systems, graphical and visual techniques (semiotic resources), etc. Anyway, the situation is very similar to natural language and other semiotic resources: from an expressive point of view they are reasonable, frequently effective and sometimes ineffective. While it seems to be indispensable, is there any alternative?

Acknowledgements The author is grateful to the Spanish Ministry of Science and Innovation for supporting this work (FFI2014-53164 and PGC2018-095941-B-I009).

References

Baber, R. L. (2011). *The language of mathematics. Utilizing math in practice.* Hoboken, NJ: John Wiley and Sons.

Bateman, J. (2014). *Text and image. A critical introduction to the visual/verbal divide*. London: Routledge.

Berger, J. (1972). *Ways of seeing*. London: BBC and Penguin Books.

Burton, L., & Morgan, C. (2000). Mathematicians writing. *Journal for Research in Mathematics Education, 31*(4), 429–453.

Cifoletti, G. (2006). Mathematics and rhetoric. Introduction. *Early Science and Medicine, 11*, 369–389.

Colyvan, M. (2009). Mathematics and the world. In A. D. Irvine (Ed.), *Philosophy of mathematics* (pp. 651–702). Oxford: Elsevier.

Davis, P. J. (1974). Visual geometry, computer graphics and theorems of perceived type. In J. P. LaSalle (Ed.), *The influence of computing on mathematical research and education* (pp. 113–127). Providence, RI: American Mathematical Society.

Descartes, R. (1682). *Renati Descartes epistolae: partim ab auctore latino sermone conscriptae, partim ex Gallico translatae: in quibus Omnis generis quaestiones philosophiae tractantur...: pars prima*. Amstelodami: Ex typographia Blauiana: sumptibus societatis, Universitat de València, Biblioteca Històrica.

Devlin, K. (1998). *The language of mathematics. Making the invisible visible.* New York: W. H. Freeman and Company.

Field, H. (1989). *Realism, mathematics and modality*. Oxford: Blackwell.

Frege, G. (1953). *The foundations of arithmetic* (2nd revised ed., J. L. Austin, Trans.). Oxford: Blackwell.

Grattan-Guinness, I. (2008). Solving Wigner's mystery: The reasonable (though perhaps limited) effectiveness of mathematics in the natural sciences. *The Mathematical Intelligencer, 30*(3), 7–17.

Halliday, M. A. K. (2004). *The collected works of M. A. K. Halliday*. Volume 5: *The Language of Science* (J. J. Webster, Ed.). London: Continuum.
Jewitt, C. (Ed.). (2009). *The Routledge handbook of multimodal analysis* (2nd ed.). London: Routledge, 2014.
Kant, I. (1786). *The metaphysical foundations of natural science* (M. Friedman, Trans. & Ed.). Cambridge, MA: Cambridge University Press, 2004.
Kvasz, L. (2008). *Patterns of change. Linguistic innovations in the development of classical mathematics*. Basel: Birkhäuser.
Morgan, C. (2014). Mathematical language. In S. Lerman (Ed.), *Encyclopedia of mathematics education* (pp. 388–391). New York: Springer.
O'Halloran, K. L. (2005). *Mathematical discourse. Language, symbolism and visual images*. London: Continuum.
O'Halloran, K. L. (2009). Historical changes in the semiotic landscape: From calculation to computation. In C. Jewitt (Ed.), *The Routledge handbook of multimodal analysis* (pp. 98–113). London: Routledge.
O'Halloran, K. L. (2015). Mathematics as multimodal semiosis. In E. Davis & P. J. Davis (Eds.), *Mathematics, substance and surmise views on the meaning and ontology of mathematics* (pp. 287–303). New York: Springer.
Panza, M., & Sereni, A. (2015). On the indispensable premises of the indispensability argument. In G. Lolli, M. Panza, & G. Venturi (Eds.), *From logic to practice. Italian studies in the philosophy of mathematics* (pp. 241–276). New York: Springer.
Pólya, G. (1954). *Mathematics and plausible reasoning* (2nd ed.). Princeton, NJ: Princeton University Press, 1968.
Pólya, G. (1963). *Mathematical methods in science*. Washington, DC: The Mathematical Association of America.
Putnam, H. (1979). *Mathematics matter and method: Philosophical papers vol. I* (2nd ed.). Cambridge, MA: Cambridge University Press.
Quine, W. V. O. (1980). *From a logical point of view* (2nd ed.). Cambridge, MA: Harvard University Press.
Quine, W. V. O. (1981). *Theories and things*. Cambridge, MA: Harvard University Press.
Sarukkai, S. (2005). Revisiting the "unreasonable effectiveness" of mathematics. *Current Science, 88*(3), 415–422.
Sfard, A. (2008). *Thinking as communicating. Human development, the growth of discourses, and mathematizing*. New York: Cambridge University Press.
Steiner, M. (1998). *The applicability of mathematics as a philosophical problem*. Cambridge, MA: Harvard University Press.

Steiner, M. (2005). Mathematics — Application and applicability. In S. Shapiro (Ed.), *The Oxford handbook of philosophy of mathematics and logic* (pp. 625–650). Oxford: Oxford University Press.

Weinberg, S. (1993). *Dreams of a final theory*. New York: Vintage.

Weyl, H. (1949). *Philosophy of mathematics and natural science* (Revised and augmented English edition based on a O. Helmer, Trans.). Princeton, NJ: Princeton University Press.

Wigner, E. P. (1960). The unreasonable effectiveness of mathematics in the natural sciences. *Communications on Pure and Applied Mathematics, 13*(1), 1–14.

9

Characterization of Scientific Prediction from Language: An Analysis of Nicholas Rescher's Proposal

Amanda Guillan

Language is one of the constituents of science, and scientific language is studied by the semantics of science. Scientific prediction can be also understood as language, so it can be analyzed within the framework of the theory of meaning. Therefore, firstly scientific prediction can be considered as a statement. From this perspective, the nexus with knowledge about the future plays a relevant role. Secondly, we can pay attention to the timing feature, which allows us, in principle, to make a distinction between a predictive statement and a retrodiction (a statement about the past).

In turn, it is possible to broaden the analysis of the language of prediction by considering the differences between the language of basic science and the language of applied science. On the one hand, in basic science there is a nexus between language and knowledge that can lead to distinguish foresight, prediction, and forecast as statements about the future with specific features. On the other hand, in applied science there is a

A. Guillan (✉)
Center for Research in Philosophy of Science and Technology,
University of A Coruña, Ferrol, Spain

W. J. Gonzalez (ed.), *Language and Scientific Research*,
https://doi.org/10.1007/978-3-030-60537-7_9

nexus between language and human activity, where decision-making is related to prescription. Finally, in the field of the application of science, predictive statements play a role in planning, so there is a linkage between language and action.

All these considerations lead to the problem of the limits of the language of prediction. In principle, it can be considered from two directions: (1) the barriers between the scientific predictive language and the non-scientific predictive statements; and (2) the confines or ceiling of the predictive language of science. In both cases, the nexus between language and knowledge has a clear relevance.

These questions can be addressed within the framework of Nicholas Rescher's system of pragmatic idealism. Certainly, he has made important contributions to the philosophical reflection on scientific prediction (Rescher 1998a). Although his perspective is mainly focused on epistemological, methodological, and ontological issues, he also pays attention to the semantic, logic, axiological, and ethical aspects related to prediction in science (see Guillan 2017). This paper is focused on the analysis of scientific prediction from language; particularly in the characterization that Rescher suggests about scientific prediction as a statement about the future.

Insofar as he gives primacy to the view of meaning as use, his starting point is also a pragmatic conception when he analyses scientific prediction from the perspective of language (see Rescher 1998b). He is interested, above all, in the process of meaning. For him, language might be seen as a "tool" for communication, and his focus is on the conditions which allow sharing messages with informative content in the communicative practice.[1]

Rescher provides a characterization of scientific prediction as a result of an activity that seeks to obtain justified answers to meaningful question about future facts and events (cf. Rescher 1998a, 37–39); a characterization which is in tune with his pragmatic conception of meaning. From this viewpoint, it should be emphasized that (a) regarding communication, use conditions are more important than truth conditions (cf.

[1] His pragmatic approach regarding language and scientific prediction has been also analyzed in Guillan (2017), ch. 2, pp. 37–65.

Rescher 1998b, 61–75); and (b) within the framework of a pragmatic philosophy, prediction takes place in an "active context" (Gonzalez 2010, 260). Within these coordinates, he maintains that, for prediction, "correctness" is more relevant than "truth" (Rescher 1998a, 70), and that predicting is an activity that provides justified and meaningful claims about the future (Rescher 1998a, 37–38).

Within this framework of the primacy of pragmatics, this paper seeks to offer an analysis of the predictive statements in Rescher's proposal. In order to do this, the problems at stake are considered. (i) The attention goes to his proposal about prediction as a statement. Thus, the features of the predictive statements are studied and the timing feature is analyzed, so the problem of retrodiction is also considered. (ii) The focus is on different types of scientific prediction, which Rescher does not develop in an explicit way. (iii) His approach regarding the limits of prediction and language is analyzed.

1 Scientific Prediction in the Framework of the Meaning Theory

From the idea of prediction as statement that is about non-observed (or unobservable) phenomena, Rescher offers four main features in order to characterize it (cf. 1998a, 54–55): (a) it is oriented towards the future; (b) it is correct or incorrect; (c) it is meaningful; and (d) it is informative. This characterization can be broaden if we highlight that prediction is related to something expected, a feature which leads to the concept of novelty. From this viewpoint, scientific prediction is connected with the notion of "novel facts" (cf. Lakatos 1978) which involve novelty in some relevant sense (ontological, epistemological, or heuristic).

1.1 Features of Scientific Prediction as a Statement

In his monograph on scientific prediction, Rescher offers a definition of prediction with a pragmatic component. According to him, "prediction (…) is our instrument for resolving our meaningful questions about the future, or at least of *endeavoring* to resolve them in a rationally cogent manner" (1998a, 39). From this perspective, prediction is the result of an activity of question-solving. As an answer to a predictive question, prediction can be considered as a statement. In this regard, Rescher considers that it has a series of characteristics (cf. 1998a, 54–55): (a) it is oriented towards the future; (b) it is correct or incorrect, rather than be true or false; (c) it is meaningful; and (d) it is informative.

The content of a predictive statement is future-oriented. The very title of his monograph on prediction (*Predicting the Future*) is significant in this regard (1998a). He highlights the timing feature, since — in his judgment — "only statements that reach beyond the facts of the past-&-present — statements that could, in principle, be falsified by yet unrealized developments — can qualify as genuinely predictive" (Rescher 1998a, 46). But not every statement which is future-oriented is, *eo ipso*, a predictive statement. In this regard, Rescher distinguishes *precognition* from *prediction* (1998a, 53–54). The former is a matter of *direct* insight into the future; while the latter requires *rational evidentiation.* There should be rational bases that support the inference from the available data to the possible future. Thus, the rational component is highlighted. In Rescher's words: "prediction (…) is a matter of thought and not perception" (1998a, 54).

As a result of a rational process of inference, the content of the predictive statement can have objectivity. It should be emphasized that, although Rescher's philosophy is a system of pragmatic idealism, he admits realist notions such as objectivity (cf. Guillan 2020a). This feature is especially relevant when scientific prediction is at stake, since we need rational bases (theoretical or empirical) that justified the predictive statement, mainly if we consider prediction as a test for the validity of theories or as a guide in the prescriptive activity of the applied sciences, as it is emphasized in different versions of scientific realism (cf. Gonzalez 2020). Moreover, "prediction can be used as a source for endorsing realism in science, insofar as successful novel predictions give grounds for objectivity in scientific research" (Gonzalez 2020, 253).

Related to his pragmatic approach to language is the second feature that Rescher assigns to predictive statements: a prediction should be valued as correct or incorrect, rather than as a true or false statement. Thus, truth conditions are in the background; while use conditions are emphasized. Insofar as a predictive statement is about a not yet realized future we cannot say now if it is true or false, but we can estimate its correctness according to its credibility, which depends on the rational bases of the prediction. In other words, the credibility and, consequently, the correctness of a predictive statement rest on the available knowledge about the past and present phenomena, and on our knowledge about how future phenomena might be.

Both characteristics (correctness and credibility) are emphasized when Rescher considers the predictive statements from an axiological viewpoint.[2] Thus, although "truth" is a central notion in his philosophy of science,[3] it is pushed into the background when prediction (as a statement about future facts and events) is considered. This feature is closely related with the third characteristic of the predictive statements pointed out above: its meaningful character. A prediction is a meaningful statement: "just as any meaningful assertion whatever makes a *statement*, so any meaningful assertion about the future makes a *prediction* in the strictly grammatical sense" (Rescher 1998a, 54–55). Again, the rational bases establish the main difference between the genuine prediction and other forms of anticipation of future such as precognition or prophecy: "only *reasonable* and *substantive* predictions (…) are of cognitive interest" (1998a, 55).

Here, the forth characteristic of the predictive statements intervenes: a genuine prediction must be informative. Informativeness is related to values such as definiteness, exactness, detail or precision (cf. Rescher 1998a, 62). In this regard, Rescher considers that there are difficulties to achieve both predictive security and informativeness. In his judgment, highly informative predictions are, in general, less secure than predictions with a low degree of informativeness. Thus, although it is desirable to achieve both highly informative and secure (credible) predictions, we usually have to seek an equilibrium that is on a moderate degree regarding both requirements.

[2] Besides correctness and credibility, Rescher points out four other values that a predictive statement should fulfill: Relevance, accuracy, precision, and robustness (1998a, 119–125). On Rescher's axiological approach for philosophy, in general, and prediction, in particular, see Guillan (2017, ch. 8).

[3] See Guillan (2017, 2020a).

Besides these characteristics, it should be pointed out — in my judgment — that, as a statement that is oriented towards the future and made on rational bases, scientific prediction has to do with something expected (cf. Gonzalez, 2010, 284). To be sure, if a prediction is *credible*, it means that there are grounded reasons in order to expect that what prediction claims will happen. This feature is related to *novelty* in some relevant sense (ontological, epistemological, and heuristic) and with the notion of *novel facts*. Even more, it has been claimed that prediction is a research on novel facts (cf. Gonzalez 2010, 11). The notion of novel facts is also relevant in order to shed light on the debate regarding the possibility of having a genuine retrodiction (i.e., a prediction of past).

1.2 The Timing Feature: Prediction and Retrodiction

There are at least four options when the notions of novelty and novel facts are considered (cf. Gonzalez 2020, 262): (1) the ontological novelty; i.e., novelty in the strictly temporal sense, so the novel facts are about something (a phenomenon or an event) that has not occurred yet; (2) epistemological novelty; i.e., the facts are novel with regard to the background theory; (3) heuristic novelty; that is, novelty with respect to design; and (4) a kind of temporal novelty for the individual.[4] To the extent that Rescher insists on prediction as a statement whose content is oriented towards the future, where the ontological dimension of novelty has primacy, he rejects de facto the possibility of a retrodiction as a prediction of past.

The possibility of a prediction of past has been maintained, among other authors, by Milton Friedman (1953) and Stephen Toulmin (cf. 1961, 26–27). According to Friedman, prediction is not necessarily about future events, it may be about phenomena that have occurred but that have not been observed yet. Even more, a prediction can be about phenomena that are not known to the predictor (cf. 1953, 9). Toulmin also admits the "prediction of past," and even the "prediction of present." For

[4] On the notion of "novel facts," see also Gonzalez (2001, 2014).

him, prediction is just an "assertion about the occurrence of a particular sort of event — whether in the past, present, or future" (1961, 31).

When the focus is on the epistemological novelty, rather than in temporal novelty, the possibility of retrodiction arises (see Feinberg et al. 1992). This is even clearer when novelty is with regard to the person who makes the prediction. From this perspective, there are still approaches to prediction that maintain that "a claim that we should find a piece of evidence in a particular context is a prediction, even if that context occurred in the past (…). Indeed, if someone claims that some event should have taken place, but they do not know that it did, and someone else already does, that should still count as a prediction" (Douglas 2009, 446).

Insofar as Rescher highlights the temporal factor, the possibility of having a prediction of past (a retrodiction), or even a prediction of present, is rejected. For him, "only statements (…) that could, in principle, be falsified by yet unrealized developments, can qualify as genuinely predictive" (Rescher 1998a, 46). Again, the truth or falsity of a predictive statement depends on the future development of the events or phenomena. He also points out that it is not possible to predict something that has already happened, only future reactions to past events can be predicted (cf. Rescher 1998a, 254, n. 66).

Although Rescher criticism is — in my judgment — right, the notion of novel facts can shed further light on this question. It is relevant both to clarify the concept of prediction and to discard the possibility of a prediction of past or a prediction of present. In effect, the content of a predictive statement refers to an unobserved or (now) unobservable phenomena or events. If the novel facts at issue in a prediction entail an ontological novelty, then the relation between the prediction and the possible future (the temporal novelty) is clear. This is the case, for example, in the prediction about the winner of an election, in an earthquake prediction, or in weather forecast.

When prediction involves a novelty in the epistemological sense, the relation between prediction and something expected is clear. Thus, the inference from the available data leads to a predictive statement, which claims that is expected that something exits or will be discovered. This is the case, for example, in the prediction of return of Halley's comet, the discovery of Neptune or the prediction about the existence of the neutrino. This feature of epistemological novelty does not justify the possibility of a genuine retrodiction (i.e., a prediction of past). On the contrary, it stresses that prediction is related with something expected. In sum, as a statement, prediction is oriented towards the possible future (the temporal factor), and claims that it might be expected that something will happen (the epistemological factor, that is clear, for example, when prediction is about the existence of an unobserved entity).

2 To Broaden the Language of Prediction in Its Nexus with Knowledge

An attempt to broaden the language of prediction should take into account the differences in the language between the different types of scientific activity: basic, applied, and of application. In basic science, predictive language is related to knowledge, so it is possible to distinguish between different kinds of predictions: foresight, prediction in the strict sense, and forecast. In applied science, the linkage is between predictive language and human activity, so prediction is the previous step to prescription. Finally, in the application of science, prediction is a guide for human action, where planning and policy should be considered.

2.1 In the Context of Basic Science: Foresight, Prediction, Forecast

Within the realm of the basic science, Rescher highlights the role of prediction as a test for scientific theories and, consequently, he considers predictive success as an indicator of scientific progress. The nexus between prediction and knowledge is then clear. In this regard, semantics of scientific prediction requires a high level of rigor that lead to distinguish between the different possible types of scientific prediction. Anastasios Brenner emphasizes this requirement when he states that "many languages distinguish, with regard to rational anticipation, between a strong sense and a weak sense (…). Furthermore, the semantic field designating the different degrees and types of foretelling is quite rich: vaticination, prognostic, warning, etc. It remains then to clarify this varied usage" (Brenner 2020, 88).

Regarding language, Rescher contemplates some distinctions that have to do with prediction. In this regard, he distinguishes qualitative and quantitative prediction, as well as generic and specific prediction (cf. 1998a, 198). In my judgment, these distinctions are not good enough in order to encompass the entire possible field. He uses equally terms such as "foresight," "prediction," and "forecast," and despite his concern about assigning a specific term to each type of prediction, he concludes that "the actual albeit regrettable fact is that English does not afford us this terminological luxury" (1998a, 55).

However, the three terms used by Rescher (foresight, prediction, and forecast) have been proposed as different types of predictions according to the degree of control of the variables that they achieve (cf. Fernández Valbuena 1990). In my viewpoint, it is desirable to achieve such a typological variety in the realm of scientific prediction, which has been already achieved in the field of scientific explanation. For this reason, Rescher's approach to the semantic features of prediction should be broadened in this direction.

Related to the degree of control of the variables concerning the future, there are differences that have to do with the phenomena studied, the methodology used, or the problem discussed. Consequently, it is possible to propose specific terms to refer to each type of prediction. A foresight entails control of the variables related to the possible future. Thus, "it is a presentation about the state of a variable within a period of time, when the variable is directly or indirectly under our control" (Fernández Valbuena 1990, 388). VAT is an example of that type of variable. It is also the most secure kind of prediction; because control over the variables at stake involves that, in principle, there will not be changes that prevent foresight for success.

A prediction, in the strict sense, takes into account variables that predictors control along with other variables that they do not control. It can be also the case that there is not a complete control of the variable because there are some factors of it that we do not control, either due to endogenous or to exogenous factors (inflation or unemployment are examples of these variables). Thus, a prediction, *sensu stricto*, is less reliable than a foresight. Finally, a forecast entails a margin of error. It is not a very precise prediction, since it establishes a margin of error where the forecasted phenomenon is expected to be placed (for example, a weather forecast).

These differences have to do with the model used for predicting. In effect, a determinist model, which does not use random variables, provides predictions, while a stochastic model, which does use random variables, provides forecasts (cf. Gonzalez 2015, 69). It seems clear that both scientific and philosophic language are (and should be) more precise than ordinary language. As A. Brenner has pointed out, ordinary languages distinguish between a strong and a weak sense of prediction (cf. 2020, 88), and this kind of distinction should be further developed within the realm of the semantics of prediction. However, there is no consensus regarding the terms to be used and their specific content, as can be seen in the fact that "prediction" and "forecast" are often used as synonymous words (Gonzalez 2010, 262n).

2.2 In the Field of Applied Science and Application of Science: From Prescription to Planning

In the realm of the applied sciences, prediction is used as a guide for action, so it is the previous step to prescription. In effect, the anticipation of the possible future (the prediction) is needed in order to propose paths of action which should been followed in order to solve a concrete problem (the prescription). This problem-solving activity is the characteristic feature of the applied sciences (see Niiniluoto 1993, 2014, 2020). Rescher's main interest is on the basic scientific activity, mainly in the realm of the natural sciences, so he rarely pays attention to this role of scientific prediction as a guide for prescription in the applied scientific research.

"Prescription," thus, must be taken into account besides "foresight," "prediction," and "forecast," when the semantics of prediction is analyzed. As prediction is chronological prior to prescription and, furthermore, it is always required in order to make prescriptions (cf. Gonzalez 2008), questions such as the objectivity, credibility, or informativeness of the predictive statement are crucial, as well as the degree of control of the variables relevant to the phenomena or event predicted. According to Simon (1990), prescription in the applied sciences is usually oriented to deal with phenomena that predictors cannot control, so the aim is usually to achieve the best possible adaptation to the phenomena predicted. From this perspective, a forecast may be good enough in order to propose an effective prescription (cf. Guillan 2017, 159–161).

Although Rescher is mainly focused in the features of prediction in basic science, the pragmatic component of his philosophical system leads him to consider how agents (individuals or groups) use predictive knowledge in order to shape their action, since — in his judgment — "to act, to plan, to survive, we must anticipate the future" (1998a, 65). This leads to another concept, "planning," which must be considered when the application of science is at stake.

The application of science, as Niiniluoto (cf. 1993, 9) has pointed out, has to do with the use that agents (individuals or groups) make of scientific knowledge and methods with the aim to solve practical problems of action. Thus, semantics of scientific prediction must take into account the concept of "planning," since predictive knowledge is used in this realm in order to guide human action, so it has a role in decision-making and policy issues (cf. Guillan 2017, 162–164).

As it happens in the realm of the applied sciences, forecasts might be good enough in order to solve the practical problems of the application of science (see Masterton 2014). Thus, for Rescher, "effective operation [of policy guidance] does indeed *not* demand categorical predictions, since even merely probabilistic considerations can provide serviceable and perfectly cogent guidance to action" (1998a, 198). This is the case of the application of economics, a realm in which the relevance of planning is highlighted by Rescher, since "*policy guidance* is one of the main aims of the macroeconomic enterprise" (1998a, 198).

However, the objectivity of the predictive content is again crucial, since objectivity "increases the possibility of finding the actual solution of the problems (which is relevant in cases such as the economic crisis of recent years or the Covid-19 pandemic)" (Gonzalez 2020, 265). In this regard, reliability and objectivity of predictions are required when prediction serves as a support in decision-making, both in professional contexts (such as medical practice) and in policy issues (such as public health or environmental policy). Here, reliable predictions are desirable since they can contribute to achieve an agreement among the experts regarding both the adequate solutions and their possible consequences.[5]

3 Towards the Limits of Language and Scientific Prediction

The problem of the limits of science can be addressed, in principle, from two different perspectives (cf. Radnitzky 1978). The first one considers the existence of "frontiers" or "barriers" (*Schranken*) of the scientific activity; i.e., the boundaries that allow us to demarcate "science" from "non-science" (or even "pseudoscience). The second one has to do with the possibility of having "confines" (*Grenzen*) in the scientific activity; that is, a ceiling that establish a terminal limit to the scientific knowledge that scientific activity can achieve.[6]

3.1 Language and the Limits of Prediction as Barriers: Scientific and Non-scientific Predictions

From the first perspective (the limits of science seen as barriers), a rigorous analysis of prediction from language should consider the difference between "scientific prediction" and "non-scientific prediction." In effect, there are both scientific and non-scientific predictive statements, and the barriers between them should be specified. Rescher does not develop this issue, since his focus is in predicting as a rational activity. In this regard, he does distinguish between "reasoned" and "unreasoned" predictions (cf. Rescher 1998a, 53–56).

On the basis of the rational character of prediction, Rescher establishes differences between a genuine "prediction" and a "precognition" (cf.

[5] This perspective of the application of science clearly links prediction with ethical issues, which arises due to the need of regulating professional practices and reducing risks in policy issues that have clear repercussions for society (see Guillan 2017, 318–321).

[6] An analysis of the limits of science from both perspectives can be found in Gonzalez (2016).

1998a, 53–54). Thus, a "precognition" is "a matter of insight into the future through a direct, non-discursive prevision into the makeup of things to come" (1998a, 53). The lack of rational evidentiation entails that a "precognition" is not credible, so it does not provide a valuable knowledge about the possible future and, consequently, it should not be used as a guide for action.[7] The same can be told about a "clairvoyance" (1998a, 54) or a "prophecy" (1998a, 55).

Realism and objectivity are, then, crucial. This feature leads Rescher to explicitly criticize Friedman's proposal of methodological instrumentalism (cf. Rescher 1998a, 109 and 194–196). For Friedman, "the only relevant test of the validity of a hypothesis is comparison of its predictions with experience" (1953, 8–9). Thus, he considers that scientific methods should be subordinated to the aim of prediction. Furthermore, economic models do not need realistic assumptions, since they are only assessed on the basis of their predictive ability (the correctness in the results).

Contrary to Friedman, Rescher maintains that "'models' that do not actually *model* — that is, do not isomorphically reflect the real world's arrangements in their own makeup — will for this very reason fail to parallel the real world's *modus operandi* and accordingly prove predictively failure prone" (1998a, 109). Moreover, in Rescher's approach, the realism of the assumptions is a sufficient criterion to differentiate between a rational prediction and a non-rational one.

Thus, he considers that "only *reasonable* and *substantive* predictions — those which are both informative and can be rendered plausible to other people by way of substantiation — are of any cognitive interest" (Rescher 1998a, 55). In this regard, it can be seen how he does admit differences regarding language. There is then a starting point to demarcate a scientific prediction from a non-scientific one: the rational bases. In effect "unreasoned predictions" are different from scientific predictions (cf. Rescher and Helmer 1959, 32). However, although being reasoned is a necessary condition for a prediction to be scientific; it is not a sufficient condition: there are also non-scientific reasoned predictions.

[7] "Predictions whose merits can be recognized only after the fact with the wisdom of retrospective hindsight are effectively useless," Rescher (1998a), p. 55.

In fact, Rescher's approach takes into account that there are two types of rational predictions. On the one hand, there are reasoned predictions based on everyday experience; and, on the other hand, there are genuine scientific predictions (cf. 1998a, 57). Both are equally fallible, since they are oriented towards a now unobservable future. However, scientific predictions are — for him — superior as science; that is, the scientific predictive statement is the result of using the scientific available knowledge and applying scientific well-established methods.

Again, it can be seen how his proposal on prediction is mainly focused on epistemological, methodological, and ontological issues. Regarding the demarcation between scientific and non-scientific predictions, he considers that the differences are basically at the methodological level, although he avoids a methodological instrumentalism. Moreover, as it has been pointed out above, he criticizes Friedman's approach in this regard.

However, in order to go deeper in the distinction between scientific and non-scientific prediction, it seems advisable to take into account the different components of prediction, which can be based on the different constituents of a science (cf. Gonzalez 2020, 261). Consequently, the problem of the demarcation between scientific and non-scientific prediction can be considered from the semantics, logic, epistemology, methodology, ontology, axiology, and ethics of prediction.

In effect, the perspective of language is relevant when the demarcation problem is considered. From this viewpoint, the specificity of the language of scientific prediction in contrast to non-scientific prediction must be taken into account. Thus, scientific language is more precise than an ordinary language (cf. Gonzalez 2020, 261), and issues such as the accuracy and precision in the sense and reference of the terms used in a scientific predictive statement are especially important to shed light on the demarcation problem.

3.2 The Limits of Scientific Prediction as "Confines:" "Non-predictability" and "Unpredictability"

From the second perspective of analysis regarding the limits of science, the confines of the predictive language of science should be analyzed. When the limits of prediction are understood as "confines" or "ceiling," a distinction is required between two different kinds of impossibility of prediction: "not predictability" and "unpredictability" (Gonzalez 2015, 56). An event or phenomenon is not predictable when there is now an impossibility of predicting it (at any level of accuracy and precision). Meanwhile, unpredictability entails a complete impossibility of predicting something (now and in the future). In this regard, Rescher's main interest is in the consequences of unpredictability and not predictability for the advancement of science.

He maintains that there are two main types of limits that affect science, in general, and scientific prediction, in particular: the epistemological and the ontological ones (see Rescher 1998a, 133–156, 1999). Epistemological limits have to do with the kind of knowledge that is achievable by scientific prediction, and they are related with agents, since predictors have limited cognitive abilities (cf. Guillan 2016a). In turn, ontological limits are related with the future-oriented nature of prediction, which deals with phenomena that are still open since they have not occurred yet (cf. Guillan 2016b). Moreover, in Rescher's approach, ontological limits have repercussions on the methodological side, so there are also methodological limitations that affect the scientific predictive processes (cf. Guillan 2020b).

Among the epistemological limits, Rescher highlights (a) ignorance (the lack of information), (b) uncertainty, and (c) inferential incapacity (see Rescher 2009, ch. 6, 91–122). Regarding the ontological limits, Rescher highlights (i) anarchy and volatility, (ii) chance, chaos and arbitrary choice, and (iii) creativity (cf. 1998a, 133–156). Unpredictability is related to anarchy. Insofar as "rational prediction pivots on the existence of some sort of appropriate *linkage* that connects our predictive claims with the input data that provide for their justification" (Rescher 1998a, 87), anarchic systems and phenomena are unpredictable: there is no a linkage between their past behavior and their future development. Thus, unpredictability has to do with the very intractability of phenomena.

Not predictability is instead related to volatility of phenomena (which has to do with the behavior of the processes over time), and with the epistemological limitations of the agents. It is also linked to the temporal projection of the prediction, since in the short run processes are usually more stable than in the medium or in the long run. In effect, predictors search for the stable elements that allow them to overcome the obstacles to predictability in order to make a better projection of the phenomenon or system into the future (cf. Gonzalez 2003). In this way, something that is not predictable now might become predictable in the future.

Consequently, there is a clear nexus between the notions of "not predictability" and "unpredictability" and two different views of the limits of science as confines: the limits of science in the weak sense and the limits in the strong sense (cf. Gonzalez 2010, 274–275). There are limits that affect science in the weak sense since there are scientific questions that we cannot answer now on the basis of the available knowledge. In turn, the limits in the strong sense are those questions that we will not answer in the future (even in the long run).

Rescher calls the limits in the strong sense the "insolubilia" problems of science (cf. 1998a, 186–188, 1999, 111–127). In this regard, he considers that it is difficult to establish now that some problems are unsolvable, because there are also difficulties to predict how future science will be (1998a, 177–183). In effect, we do not know now what we will know in the future. Regarding scientific prediction, this issue involves that there are also difficulties in order to say that a concrete phenomenon or system is unpredictable.

Thus, although anarchic phenomena are (due to their very nature) unpredictable, Rescher takes a wary attitude regarding unpredictability: on the basis of the available knowledge we can say that a phenomenon is anarchic, but our knowledge is fallible and can change in the future (Rescher, *Personal communication*, July 1, 2014). This issue is assumed in recent approaches to the problem of unpredictability: "Some things that seem unpredictable may actually be more predictable than we think using the right empirical tools. As we expand our notion of what is predictable, new applications will arise" (Kleinberg et al. 2015, 494).

At the same time, Rescher maintains that achieving a complete predictability is not possible. Insofar as science is *our* science, the ideal of a perfect science is unattainable (cf. Rescher 1999, 2012). In this regard, he highlights the epistemological and ontological limits of scientific prediction. However, as it happens in the case of the limits of predictions as "barriers," there are confines to the scientific predictive activity related with the different constituents of science. Thus, obstacles to scientific prediction can be analyzed from the semantic, logical, epistemological, methodological, ontological, axiological, and ethical viewpoints (cf. Gonzalez 2010, 277–281). This means that there are limits to science related to language, the structure of the scientific theories, the kind of knowledge, the processes used, the reality researched, and the values that intervene (among them, ethical values).

Furthermore, it can be said the first limit to knowledge is rooted in language, because it is difficult to know something that we cannot state. Properly, the limits to scientific prediction related to language have to do with "the difficulties to identify new phenomena — their sense and reference" (Gonzalez 2010, 275). In effect, the advancement of science requires words that involve accurate and precise sense and reference. Insofar as Rescher maintains a pragmatic approach regarding language, he does not develop this kind of issues, so it is possible to broaden his approach in this direction.

Acknowledgements This paper is related to the research project FFI2016-79728-P supported by the Spanish Ministry of Economics, Industry and Competitiveness (AEI).

References

Brenner, A. (2020). Evolving realities: Scientific prediction and objectivity from the perspective of historical epistemology. In W. J. Gonzalez (Ed.), *New approaches to scientific realism* (pp. 87–103). Berlin/Boston: De Gruyter.

Douglas, H. E. (2009). Reintroducing prediction to explanation. *Philosophy of Science, 76*(4), 444–463.

Feinberg, G., Lavine, S., & Albert, D. (1992). Knowledge of the past and future. *The Journal of Philosophy, 89*(12), 607–642.

Fernández Valbuena, S. (1990). Predicción y Economía. In W. J. Gonzalez (Ed.), *Aspectos metodológicos de la investigación científica* (2nd ed., pp. 385–405). Madrid-Murcia: Ediciones Universidad Autónoma de Madrid and Publicaciones Universidad de Murcia.

Friedman, M. (1953). The methodology of positive economics. In M. Friedman (Ed.), *Essays in positive economics* (pp. 3–43). Chicago, IL: The University of Chicago Press. (6th reprint, 1969).

Gonzalez, W. J. (2001). Lakatos's approach on prediction and novel facts. *Theoria, 16*(3), 499–518.

Gonzalez, W. J. (2003). Racionalidad y Economía: De la racionalidad de la Economía como Ciencia a la racionalidad de los agentes económicos. In W. J. Gonzalez (Ed.), *Racionalidad, historicidad y predicción en Herbert A. Simon* (pp. 65–96). A Coruña: Netbiblo.

Gonzalez, W. J. (2008). Rationality and prediction in the sciences of the artificial. In M. C. Galavotti, R. Scazzieri, & P. Suppes (Eds.), *Reasoning, rationality, and probability* (pp. 165–186). Standford: CSLI Publications.

Gonzalez, W. J. (2010). *La predicción científica: Concepciones filosófico-metodológicas desde H. Reichenbach a N. Rescher*. Barcelona: Montesinos.

Gonzalez, W. J. (2014). The evolution of Lakatos's repercussion on the methodology of economics. *HOPOS: The Journal of the International Society for the History of Philosophy of Science, 4*(1), 1–25.

Gonzalez, W. J. (2015). *Philosophico-methodological analysis of prediction and its role in economics*. Dordrecht: Springer.

Gonzalez, W. J. (2016). Rethinking the limits of science: From the difficulties for the frontiers to the concern on the confines. In W. J. Gonzalez (Ed.), *The limits of science: An analysis from "barriers" to "confines."* Poznan Studies in the Philosophy of the Sciences and the Humanities (pp. 3–30). Leiden/Boston: Brill/Rodopi.

Gonzalez, W. J. (2020). Pragmatic realism and scientific prediction: The role of complexity. In W. J. Gonzalez (Ed.), *New approaches to scientific realism* (pp. 251–287). Berlin/Boston: de Gruyter.

Guillan, A. (2016a). The limits of future knowledge: An analysis of Nicholas Rescher's epistemological approach. In W. J. Gonzalez (Ed), *The limits of science. An analysis from "barriers" to "confines."* Poznan Studies in the Philosophy of the Sciences and the Humanities (pp. 134–149). Leiden/Boston: Brill/Rodopi.

Guillan, A. (2016b). The obstacles to scientific prediction: An analysis of the limits of predictability from the ontology of science. In W. J. Gonzalez (Ed.), *The limits of science. An analysis from "barriers" to "confines."* Poznan Studies in the Philosophy of the Sciences and the Humanities (pp. 183–206). Leiden/Boston: Brill/Rodopi.

Guillan, A. (2017). *Pragmatic idealism and scientific prediction: A philosophical system and its approach to prediction in science.* Dordrecht: Springer.

Guillan, A. (2020a). Realistic components in the conception of pragmatic idealism: The role of objectivity and the notion of "fact". In W. J. Gonzalez (Ed.), *New approaches to scientific realism* (pp. 331–347). Berlin/Boston: de Gruyter.

Guillan, A. (2020b). Methodological incidence of the realms of reality: Prediction and complexity. In W. J. Gonzalez (Ed.), *Methodological prospects for scientific research: From pragmatism to pluralism* (pp. 81–96). Synthese Library, Springer.

Kleinberg, J., Ludwig, J., Mullainathan, S., & Obermeyer, Z. (2015). Prediction policy problems. *The American Economic Review, 105*(5), 491–495.

Lakatos, I. (1978). *The methodology of scientific research programmes.* Philosophical papers, vol. 1. Cambridge: Cambridge University Press.

Masterton, G. (2014). What to do with a forecast? *Synthese, 191*(8), 1881–1907.

Niiniluoto, I. (1993). The aim and structure of applied research. *Erkenntnis, 38*(1), 1–21.

Niiniluoto, I. (2014). Values in design sciences. *Studies in History and Philosophy of Science, 46*, 11–15.

Niiniluoto, I. (2020). Interdisciplinarity form the perspective of critical scientific realism. In W. J. Gonzalez (Ed.), *New approaches to scientific realism* (pp. 231–250). Berlin/Boston: de Gruyter.

Radnitzky, G. (1978). The boundaries of science and technology. In *The search for absolute values in a changing world. Proceedings of the VIth international conference on the unity of sciences* (Vol. II, pp. 1007–1036). New York: International Cultural Foundation Press.

Rescher, N. (1998a). *Predicting the future. An introduction to the theory of forecasting.* New York: State University of New York Press.

Rescher, N. (1998b). *Communicative pragmatism and other philosophical essays on language.* Lanham, MD: Rowman and Littlefield.

Rescher, N. (1999). *The limits of science.* Revised ed. Pittsburgh, PA: University of Pittsburgh Press.

Rescher, N. (2009). *Pragmatism. The restoration of its scientific roots.* New Brunswick: Transaction Publishers.

Rescher, N. (2012). The problem of future knowledge. *Mind and Society, 11*(2), 149–163.
Rescher, N., & Helmer, O. (1959). On the epistemology of the inexact sciences. *Management Sciences, 6*, 25–52.
Simon, H. A. (1990). Prediction and prescription in systems modeling. *Operations Research, 38*, 7–14. Compiled in Simon, H. A. (1997). *Models of bounded rationality. Vol. 3: Empirically grounded economic reason* (pp. 115–128). Cambridge, MA: The MIT Press.
Toulmin, S. (1961). *Foresight and understanding*. Bloomington, IN: Indiana University Press.

Index of Names

Note: Page numbers followed by 'f' refers to footnotes.

W. J. Gonzalez (ed.), *Language and Scientific Research*,
https://doi.org/10.1007/978-3-030-60537-7

Subject Index

A

B

Note: Page numbers followed by 'n' refers to notes

W. J. Gonzalez (ed.), *Language and Scientific Research*,
https://doi.org/10.1007/978-3-030-60537-7

U

V

W

The manufacturer's authorised representative in the EU is Springer Nature Customer Service Centre GmbH, Europaplatz 3, 69115 Heidelberg, Germany. If you have any concerns regarding our products, please contact ProductSafety@springernature.com

Printed and bound by CPI Group (UK) Ltd, Croydon, CR0 4YY

30/09/2026

02225457-0011